Radio Systems Technology

Radio Systems Technology

D C Green
MTech, CEng, MIEE

Longman
Scientific &
Technical

Copublished in the United States with
John Wiley & Sons, Inc., New York

Longman Scientific & Technical,
Longman Group UK Limited,
Longman House, Burnt Mill, Harlow,
Essex CM20 2JE, England
and Associated Companies throughout the world

Copublished in the United States with
John Wiley & Sons, Inc., 605 Third Avenue, New York, NY 10158

First published 1990
Second impression 1992

British Library Cataloguing in Publication Data
Green, D.C. (Derek Charles), *1931 –*
    Radio systems technology.
    1. Radio engineering
    I. Title
    621.384

ISBN 0-582-02697-0

Library of Congress Cataloging-in-Publication Data
Green, D. C. (Derek Charles)
    Radio systems technology/D.C. Green.
        p.   cm.
    ISBN 0-470-21672-7
    1. Radio.   I. Title.
    TK6550.G84   1990
    621.384—dc20   90–5627
                                                                        CIP

ISBN 0-470-21672-7 (USA only)

Produced by Longman Singapore Publishers (Pte) Ltd.
Printed in Singapore

Contents

Preface

This book has been written to provide a comprehensive introduction to modern radiocommunication systems. An attempt has been made throughout to keep the treatment of each topic at a level appropriate for a UK Higher Certificate/Diploma student or for an early-year degree student. This means that the reader is assumed to possess a prior knowledge of electrical principles, electronics, mathematics and radio/transmission principles of at least the standard reached by the Level III Business and Technician Education Council (BTEC) units.

The book discusses the principles of analogue modulation in Chapters 1 and 2; but gives only a brief mention of digital modulation since, at present, its radio applications are restricted to some microwave systems. Transmission lines and waveguides are important components in any radio system since they are used both to provide feeders and to simulate components and tuned circuits. Lines and waveguides are the subjects of Chapters 3 and 4. The important topic of noise is considered in Chapter 5 before the next two chapters deal with aerials. The basic concepts of aerials are introduced to the reader in Chapter 6 before some of the more commonly employed aerials are discussed in Chapter 7. Chapter 8 deals with the propagation of radio waves from one aerial to another. The treatment of communication radio receivers, in Chapter 9, assumes throughout a previous knowledge of radio receivers up to the BTEC Level III standard. No circuits are given because of demands on space, but a mention is made of some of the Plessey ICs that can be usefully employed in modern receivers. Lastly, Chapter 10 provides an introduction to microwave radio-relay systems, both terrestrial and satellite, and to land-mobile systems.

Many worked examples have been provided throughout the book to illustrate various principles and at the end of the book will be found exercises together with answers to the numerical exercises.

I wish to express my thanks and gratitude to Eddystone Radio Ltd for their permission to include the block diagrams of their 1650 and 1995 radio receivers, and to the Institution of British Telecommunication Engineers for their permission to use several diagrams from their journal *British Telecommunication Engineering*.

DCG

1 Amplitude Modulation

Some form of modulation is always employed in a radio system to frequency shift, or translate, a baseband signal from its original frequency bandwidth to a specified part of the radio-frequency spectrum. The various radio services have each been allocated particular frequency bands by international agreement (see Transmission Principles for Technicians). Different stations operating in an allocated frequency band must, of course, be positioned at different points within that band so that a wanted signal can be selected from all those signals that are simultaneously present at the receiving aerial.

Analogue modulation consists essentially of using the baseband, or modulating, signal to vary *one* of the three variables of a sinusoidal *carrier wave*. If the instantaneous voltage v_c of the carrier wave is

$$v_c = V_c \sin (\omega_c t + \theta) \tag{1.1}$$

then: (*a*) the amplitude V_c of the carrier may be varied, this is *amplitude modulation*; (*b*) the frequency $\omega_c/2\pi$ may be varied, this is *frequency modulation*; and (*c*) the phase θ can be varied to give *phase modulation*. It will be seen in Chapter 2 that frequency and phase modulation are very similar processes and they are often referred to jointly as *angle modulation*. Amplitude modulation is still the most widely used modulation method with both double-sideband and single-sideband versions being common.

Double-sideband Amplitude Modulation

For a sinusoidal carrier wave to be amplitude modulated the amplitude of the carrier must be varied in the same way as the instantaneous voltage of the modulating signal. If the modulating voltage is represented by $V_m(t)$ then the instantaneous voltage of the modulated carrier wave is

$$v_c = [V_c + V_m(t)] \sin \omega_c t \tag{1.2}$$

in which the phase angle θ of the unmodulated carrier is assumed, for convenience, to be zero.

When the modulating signal is a sine wave, i.e. $V_m(t) = V_m \sin \omega_m t$, then equation (1.2) becomes

$$v_c = [V_c + V_m \sin \omega_m t] \sin \omega_c t. \tag{1.3}$$

This equation can be expanded, using the trigonometric identity $2 \sin A \sin B = \cos (A-B) - \cos (A + B)$, to give

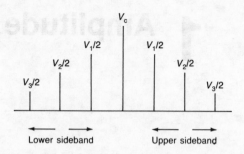

Fig. 1.1 Amplitude-modulation sidebands.

$$v_c = V_c \sin \omega_c t + V_m/2 \cos (\omega_c - \omega_m)t$$
$$- V_m/2 \cos (\omega_c + \omega_m)t. \tag{1.4}$$

Clearly, the sinusoidally modulated carrier wave contains components at three different frequencies: (a) the lower side-frequency $f_c - f_m$, (b) the carrier frequency f_c, and (c) the upper side-frequency $f_c + f_m$. The modulating signal frequency f_m is *not* present. When the modulating signal $V_m(t)$ is not of sinusoidal waveform Fourier analysis will show that it contains the sum of components at a fundamental frequency, plus one, or more, components at other, higher, frequencies. Each of these component frequencies will modulate the carrier to produce corresponding lower and upper side-frequencies in the modulated waveform. Suppose, for example, that $V_m(t) = V_1 \sin \omega_1 t + V_2 \sin \omega_2 t + V_3 \sin \omega_3 t$. Then, the instantaneous voltage of the modulated wave is

$$v_c = [V_c + V_1 \sin \omega_1 t + V_2 \sin \omega_2 t + V_3 \sin \omega_3 t] \sin \omega_c t. \tag{1.5}$$

Equation (1.5) can also be expanded to give

$$v_c = V_c \sin \omega_c t + V_1/2 \cos (\omega_c - \omega_1)t$$
$$+ V_2/2 \cos (\omega_c - \omega_2)t$$
$$+ V_3/2 \cos (\omega_c - \omega_3)t$$
$$- V_1/2 \cos (\omega_c + \omega_1)t$$
$$- V_2/2 \cos (\omega_c + \omega_2)t$$
$$- V_3/2 \cos (\omega_c + \omega_3)t \tag{1.5a}$$

which shows that the modulated wave now contains both lower and upper *sidebands* that are symmetrically situated either side of the carrier frequency. This is illustrated by Fig. 1.1.

Modulation Factor

The modulation factor m of an amplitude-modulated wave expresses the degree to which the amplitude of the carrier is varied from its unmodulated value. It is given by equation (1.6), i.e.

$$m = \frac{\text{r.m.s. value of } V_m(t)}{\text{r.m.s. value of unmodulated carrier}}. \tag{1.6}$$

When *m* is expressed as a percentage it is generally known as the *depth of modulation*. The depth of modulation must never be allowed to become greater than 100% because this situation would result in excessive distortion of the modulation envelope. For a sinusoidally modulated wave

$$m = (V_m/\sqrt{2})/(V_c/\sqrt{2}) = V_m/V_c, \text{ or from equation (1.3)}$$

$$m = \frac{V_m}{V_c} = \frac{(V_c + V_m) - (V_c - V_m)}{(V_c + V_m) + (V_c - V_m)}$$

$$= \frac{\text{maximum voltage} - \text{minimum voltage}}{\text{maximum voltage} + \text{minimum voltage}}. \quad (1.7)$$

Example 1.1

A carrier wave $v_c = 10 \sin (8 \times 10^6 t)$ is amplitude modulated by the signal $4 \sin (2 \times 10^3 t) + 1 \cos (6 \times 10^3 t)$ volts. Calculate the depth of modulation.

Solution
The r.m.s. value of the modulating signal voltage is

$$\sqrt{\left(\frac{4^2 + 1^2}{2}\right)} = 2.916 \text{ V},$$

and hence

$$m = \frac{2.916}{10/\sqrt{2}} = 0.412, \text{ or } 41.2\%. \quad (Ans.)$$

The maximum depth of modulation that a practical amplitude modulator is able to produce without generating distortion in excess of a specified limit is generally restricted. Sometimes a carrier-cancellation technique is employed to reduce the voltage of the carrier component and in this way increase the depth of modulation.

The expressions for the instantaneous voltage of an amplitude-modulated wave can be rewritten in terms of the modulation factor.

(*a*) For sinusoidal modulation

$$v_c = V_c[1 + m \sin \omega_m t] \sin \omega_c t. \quad (1.8)$$

(*b*) For three-tone modulation

$$v_c = V_c[1 + m_1 \sin \omega_1 t + m_2 \sin \omega_2 t + m_3 \sin \omega_3 t] \sin \omega_c t. \quad (1.9)$$

(*c*) For a general complex modulating signal $V_m(t)$

$$v_c = V_c\left[1 + \frac{V_m(t)}{V_c}\right] \sin \omega_c t. \quad (1.10)$$

The Root-mean-square Value of an Amplitude-modulated Wave

The r.m.s. value of an amplitude-modulated wave is the square root of the sum of the squares of the r.m.s. values of each of its component frequencies. Thus, for a sinusoidally modulated wave

$$V = \sqrt{\left[\left(\frac{V_c}{\sqrt{2}}\right)^2 + \left(\frac{mV_c}{2\sqrt{2}}\right)^2 + \left(\frac{mV_c}{2\sqrt{2}}\right)^2\right]}$$

$$= \sqrt{\left[\frac{V_c^2}{2}\left(1 + \frac{m^2}{2}\right)\right]}, \text{ i.e.}$$

$$V = \text{r.m.s. carrier voltage } \sqrt{(1 + m^2/2)}. \tag{1.11}$$

For a three-tone modulated wave,

$$V = \sqrt{\left[\left(\frac{V_c}{\sqrt{2}}\right)^2 + \frac{m_1^2 V_c^2}{4} + \frac{m_2^2 V_c^2}{4} + \frac{m_3^2 V_c^2}{4}\right]}, \text{ or}$$

$$V = \text{r.m.s. carrier voltage } \sqrt{\left(1 + \frac{m_1^2}{2} + \frac{m_2^2}{2} + \frac{m_3^2}{2}\right)},$$

i.e. $V = \text{r.m.s. carrier voltage } \sqrt{(1 + m_T^2/2)}$ (1.12)

where $m_T = \sqrt{(m_1^2 + m_2^2 + m_3^2)}$.

Example 1.2

A carrier has an r.m.s. voltage of 10 V. It is amplitude modulated by a signal having components at frequencies f_1 and f_2 when its r.m.s. voltage rises to 11.5 V. If the depth of modulation due to one of the components is 60% calculate the depth of modulation caused by the other component.

Solution

From equation (1.12), $11.5 = 10\sqrt{(1 + m_T^2/2)}$

$1.3225 = 1 + m_T^2/2$, or $m_T = 0.803$.

Therefore

$$m_2 = \sqrt{(0.803^2 - 0.6^2)} = 0.534 \text{ or } 53.4\%. \quad (Ans.)$$

If the modulating signal is of rectangular shape the number of its frequency components is very large and an alternative approach will give a simpler and more accurate result. Suppose that the rectangular waveform has a peak value of V_m volts. The maximum value of the modulated carrier will be $V_c + V_m$ volts and the minimum value will be $V_c - V_m$ volts (see Fig. 1.2). The r.m.s. value of the modulated waveform is

$$V = \sqrt{\left(\frac{1}{T}\int_0^T v^2 dt\right)} = \frac{V_c}{\sqrt{2}}\sqrt{\left(1 + \frac{m_T^2}{2}\right)} \tag{1.13}$$

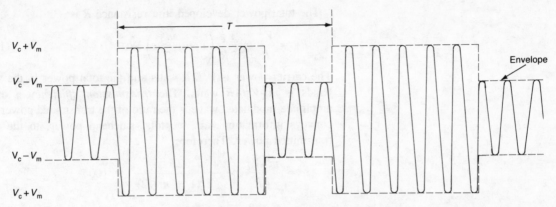

Fig. 1.2 Carrier amplitude modulated by a rectangular signal.

Example 1.3

An 8 V peak carrier wave is amplitude modulated by a square waveform of peak value 5 V. Calculate (*a*) the r.m.s. value of the modulated waveform, and (*b*) its depth of modulation.

Solution

(*a*) The maximum value of the modulated wave is 13 V and the minimum value is 3 V. Hence

$$V = \sqrt{\left[\frac{1}{T} \left(\int_0^{T/2} 13^2 \sin^2 \omega_c t \, \mathrm{d}t + \int_{T/2}^T 3^2 \sin^2 \omega_c t \, \mathrm{d}t \right) \right]}$$

$$= \sqrt{\left[\frac{1}{T} \int_0^{T/2} \frac{169}{2} (1 - \cos 2\omega_c t) \, \mathrm{d}t \right.}$$

$$\left. + \int_{T/2}^T \frac{9}{2} (1 - \cos 2\omega_c t) \, \mathrm{d}t \right]$$

$$= \sqrt{\left[\frac{1}{T} \left(\int_0^{T/2} 84.5 \, \mathrm{d}t + \int_{T/2}^T 4.5 \, \mathrm{d}t \right) \right]},$$

(since the mean value of $\cos 2\omega_c t$ over half a cycle is zero)

$$= \sqrt{\left\{ \frac{1}{T} \left[84.5 \times \frac{T}{2} + 4.5(T - T/2) \right] \right\}}$$

$$= \sqrt{44.5} = 6.67 \text{ V.} \quad (Ans.)$$

(*b*) $6.67 = \dfrac{8}{\sqrt{2}} \sqrt{\left(1 + \dfrac{m_T^2}{2} \right)}$

or $m_T = 0.884 = 88.4\%.$ (*Ans.*)

Power Contained in an Amplitude-modulated Wave

The power developed by an amplitude-modulated wave is the sum of the powers developed by the carrier component and by each of the other components in both the lower and the upper sidebands.

The total power developed in a resistance R is

$$P_T = \frac{V^2}{R} = \frac{V_c^2}{2R} \left(1 + \frac{m_T^2}{2} \right) \text{ W.} \tag{1.14}$$

The carrier power is $V_c^2/2R$ watts and the total power in the side-bands is $m_T^2 V_c^2/4R$ watts. The *transmission efficiency* η of an amplitude-modulated wave is the ratio of the transmitted power that conveys information, i.e. the total sideband power, to the total transmitted power. Therefore,

$$\eta = \frac{m_T^2 V_c^2}{4R} \times \frac{2R}{V_c^2(1 + m_T^2/2)} \times 100\%$$

$$= \frac{m_T^2}{2 + m_T^2} \times 100\%. \tag{1.15}$$

The maximum value for the modulation factor m_T is unity and then $\eta = 33.3\%$. For any other value of m the sideband power will be an even smaller percentage of the total power. This means that d.s.b.a.m. is not a very efficient method of transmitting information from one point to another. On the other hand, d.s.b.a.m. can be demodulated by a relatively simple envelope detector that produces an output voltage which is proportional to the modulation envelope. Perhaps the simplest, and the most common, version of this is the diode detector.

Example 1.4

A 10 kW carrier wave is amplitude modulated to a depth of 70%. Calculate the total sideband power and determine what percentage of the total power it is.

Solution
From equation (1.14), $P_T = 10$ kW $(1 + 0.7^2/2) = 12.45$ kW. Hence, the sideband power is $12.45 - 10 = 2.45$ kW.   (*Ans.*)
Expressed as a percentage of the total power $= 19.68\%$.   (*Ans.*)

Phasor Representation of an Amplitude-modulated Wave

The frequency spectrum of a sinusoidally-modulated d.s.b.a.m. wave contains components at the carrier frequency f_c, and at the lower, and upper side-frequencies, $f_c \pm f_m$. Equation (1.4) can be rewritten as

$$\begin{aligned} v_c = \ & V_c \sin \omega_c t + V_m/2 \sin \left[(\omega_c - \omega_m)t + \pi/2 \right] \\ & + V_m/2 \sin \left[(\omega_c + \omega_m)t - \pi/2 \right] \end{aligned} \tag{1.16}$$

and this allows the spectrum to be represented by a phasor diagram, Fig. 1.3(a). There are three phasors, one for each component, whose length is proportional to the voltage. Each phasor rotates in the anti-clockwise direction at its own angular velocity. The instantaneous envelope of the modulated wave is given by the phasor sum of the

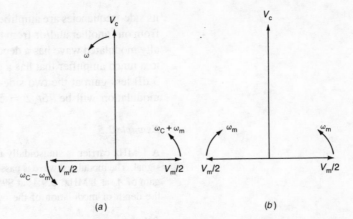

Fig. 1.3 Phasor representation of a d.s.b.a.m. wave.

(a)

(b)

three phasors; in Fig. 1.3(a) the two side-frequency phasors are in anti-phase with one another and so mutually cancel out. Consequently, the instantaneous modulation envelope is equal to the unmodulated carrier voltage. To obtain the envelope over one cycle of the modulating signal it is more convenient to take the carrier phasor as the reference and assume it to be both stationary and in the vertical plane. This is shown by Fig. 1.3(b), the two side-frequency phasors will then rotate in opposite directions with angular velocity ω_m.

Figure 1.4 shows the positions of the side-frequency phasors, and of the phasor sum of all three phasors, for intervals of $T/8$, where T is the periodic time of the modulating signal. For each time interval the phasor sum gives the instantaneous value, represented by R, of the modulated wave; joining these together gives the positive envelope of the wave. Obviously, the negative envelope can similarly be obtained.

Distortion of the Amplitude-modulated Wave

A d.s.b.a.m. waveform will suffer a reduction in its modulation depth and/or distortion if, in its transmission through a network or system,

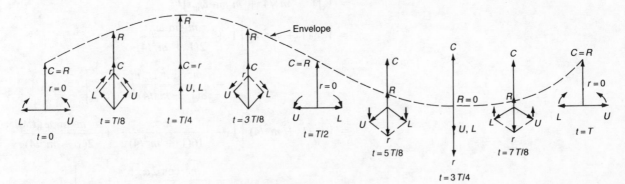

Fig. 1.4 Phasor diagram of a d.s.b.a.m. wave over one complete cycle of the modulating signal.

its side-frequencies are amplified, or attenuated, by a different amount from one another and/or from the carrier. If, for example, a sinusoidally modulated wave has a depth of modulation of 75% and is applied to a tuned amplifier that has a gain of A_v at the carrier frequency and 3 dB less gain at the two side-frequencies, then the output depth of modulation will be $75/\sqrt{2} = 53\%$.

Example 1.5

A 1 MHz carrier is sinusoidally modulated to a depth of 80% by a 5 kHz signal. The modulated wave is passed through an amplifier which has a voltage gain of A_v at 1 MHz, $0.9A_v$ at 995 kHz, and $0.8A_v$ at 1005 kHz. Calculate the depth of modulation of the output waveform.

Solution

At the output of the amplifier: carrier voltage = V_c, lower side-frequency voltage = $0.9 \times 0.4V_c = 0.36V_c$, and upper side-frequency voltage = $0.8 \times 0.4V_c = 0.32V_c$. The r.m.s. voltage

$$V = \sqrt{\left[\left(\frac{V_c}{\sqrt{2}}\right)^2 + \left(\frac{0.36V_c}{\sqrt{2}}\right)^2 + \left(\frac{0.32V_c}{\sqrt{2}}\right)^2\right]}$$

$$= \frac{V_c}{\sqrt{2}}\sqrt{[1 + 0.1296 + 0.1024]} = \sqrt{(1 + 0.232)}.$$

Hence, $m^2/2 = 0.232$ or $m = 68.1\%$.   (*Ans.*)

The calculation of the percentage distortion is more difficult. To simplify the algebra, suppose that the upper side-frequency is completely suppressed. Then the modulated wave is given by

$$v_c = V_c \sin \omega_c t + mV_c/2 \cos (\omega_c - \omega_m)t$$
$$= V_c \sin \omega_c t[1 + m/2 \sin \omega_m t] + mV_c/2 \cos \omega_c t \cos \omega_m t.$$

This has an envelope given by

$$\sqrt{[V_c^2(1 + m/2 \sin \omega_m t)^2 + (mV_c/2 \cos \omega_m t)^2]}$$

$$= V_c[1 + m \sin \omega_m t + m^2/4 \sin^2 \omega_m t + m^2/4 \cos^2 \omega_m t]^{1/2}$$

$$= V_c[1 + m^2/4 + m \sin \omega_m t]^{1/2}$$

$$= V_c \sqrt{(1 + m^2/4)} \left[1 + \frac{m \sin \omega_m t}{2(1 + m^2/4)} \right.$$

$$\left. - \frac{m^2 \sin^2 \omega_m t}{8(1 + m^2/4)^2} + \cdots \right]$$

$$= V_c \sqrt{(1 + m^2/4)} \left[1 - \frac{m^2}{16(1 + m^2/4)^2} + \frac{m \sin \omega_m t}{2(1 + m^2/4)} \right.$$

$$\left. + \frac{m^2 \cos 2\omega_m t}{16(1 + m^2/4)^2} + \cdots \right].$$

The percentage second harmonic distortion is equal to

$$\frac{m^2}{16(1 + m^2/4)^2} \times \frac{2(1 + m^2/4)}{m} = \frac{m}{8(1 + m^2/4)}. \quad (1.17)$$

Example 1.6

A carrier wave is sinusoidally modulated to a depth of 30% and has one of its side-frequencies suppressed. Calculate the percentage second-harmonic distortion of the modulation envelope.

Solution

From equation (1.17), the second-harmonic distortion

$$= \frac{0.3}{8(1 + 0.3^2/4)} \times 100 = 3.67\%. \quad (Ans.)$$

Amplitude Modulators

Probably the most commonly employed method of generating a d.s.b.a.m. wave in a radio transmitter is the anode-, or collector-modulated Class C r.f. tuned amplifier (see *Radio Systems for Technicians*). Other d.s.b. modulators utilize the non-linear relationship between the applied voltage and the resulting current of many electronic devices. Essentially, there are two types of non-linear characteristic: those in which the characteristic is continuous and can be described by a power series of the form $i = av + bv^2 + cv^3 + \ldots$, and those in which the device acts as an electronic switch (this means that the carrier voltage must be large enough, 2 V or more usually, to turn the device ON and OFF).

If the carrier wave $V_c \sin \omega_c t$ and a modulating signal $V_m(t)$ are applied in series to a non-linear device and the carrier voltage is *not* large to switch the device, then the current flowing will be given by

$$i = a[V_c \sin \omega_c t + V_m(t)] + b[V_c \sin \omega_c t + V_m(t)]^2 + c[V_c \sin \omega_c t + V_m(t)]^3, \text{ etc.}$$

The squared term will produce

$$i = bV_c^2 \sin^2 \omega_c t + bV_m^2(t) + 2bV_c \sin \omega_c t V_m(t)$$

and if a filter can be employed to pass only the terms

$$aV_c \sin \omega_c t + 2bV_c V_m(t) \sin \omega_c t,$$
$$\text{or} \quad aV_c[1 + 2b/a\, V_m(t)] \sin \omega_c t,$$

a d.s.b.a.m. wave will have been obtained. Unless the total input voltage is fairly small the cubic term cv^3 (and perhaps even higher terms) may also make a contribution to the filtered output signal and so cause distortion. Suppose, for example, that $V_m(t) = V_1 \sin \omega_1 t + V_2 \sin \omega_2 t$, then the cubic term gives $cV_m^2(t)\, V_c \sin \omega_c t$ which, when expanded, shows the presence of components at frequencies ω_c, $\omega_c \pm 2\omega_1$, $\omega_c \pm 2\omega_2$, and $\omega_c \pm (\omega_1 \pm \omega_2)$. These components may fall within the pass-band of the wanted modulating signal and then they cannot be removed by the filter. To avoid this effect it will often be found necessary to use two non-linear devices in a balanced arrangement, with respect to the carrier, so that the intermodulation terms cancel out.

The switching-type d.s.b. modulator includes one, or more, transistors that are turned ON and OFF by the carrier voltage and which effectively multiply the modulating signal by a square wave whose amplitude is either ±1 V or 0 V and 1 V. The Fourier series for a ±1 V square wave is

$$v = 4/\pi(\sin \omega_c t + 1/3 \sin 3\omega_c t + 1/5 \sin 5\omega_c t + \ldots)$$

and for a square wave whose voltage is either 0 V or 1 V is

$$v = 1/2 + 2/\pi(\sin \omega_c t + 1/3 \sin 3\omega_c t + 1/5 \sin 5\omega_c t + \ldots).$$

If the input signal is $V_B + V_m \sin \omega_m t$ then, in either case, the output voltage will be equal to $K \sin \omega_c t + K \sin \omega_c t \sin \omega_m t + \ldots$.

When an integrated circuit modulator is used, some form of *transconductance multiplier* is often employed. Figure 1.5 shows the

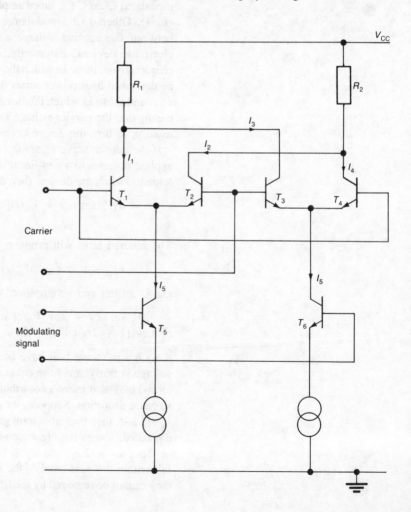

Fig. 1.5 Transconductance multiplier a.m. modulator.

basic arrangement of such a circuit. Two differential amplifiers are connected as shown and are driven by the carrier signal. The two collector load resistors are externally connected to the appropriate package pins and usually $R_1 = R_2 = R$. The differential amplifiers have individual constant-current generators to supply their *equal* d.c. emitter currents I_5. In the absence of any input modulating signal, or carrier, voltages the current relationships are: $I_5 = I_1 + I_2 = I_3 + I_4$. The carrier voltage varies the mutual conductance of transistors T_1, T_2, T_3 and T_4 so that $I_1 + I_3 = I_5 + g_{m_1}V_c$ and $I_2 + I_4 = I_5 - g_{m_1}V_c$. The differential output current is $(I_1 + I_3) - (I_2 + I_4) = 2g_{m_1}V_c$.

When the modulating signal is applied to the circuit it varies the emitter currents of the two differential amplifiers by $g_{m_2}V_m$. Now, the differential output current is $2g_{m_1}g_{m_2}V_cV_m$ and so the differential output voltage is $2g_{m_1}g_{m_2}RV_cV_m$ volts.

If the carrier voltage is small the circuit acts as a linear multiplier to produce an output proportional to $V_c \sin \omega_c t V_m(t)$. If the carrier voltage is greater than about 0.7 V it turns transistors T_1/T_4 and T_2/T_3 alternatively ON and OFF so that the differential part of the circuit switches the emitter current between the two output terminals.

Detection of an Amplitude-modulated Wave

The methods used to detect, or demodulate, a d.s.b.a.m. signal fall into one of three main classes: these are non-coherent or *envelope detection*, coherent or *synchronous detection*, and *non-linear detection*.

Envelope Detection

Since the envelope of a d.s.b.a.m. signal has the same waveshape as the original modulating signal, detection can be achieved by rectifying the envelope. The basic circuit of an envelope, or diode, detector is shown in Fig. 1.6. The input modulated signal should be of sufficiently large amplitude (about 1 V) to ensure that operation is on the linear part of the diode characteristic. Provided the time constant, C_1R_1 seconds, of the resistive load and shunt capacitor is long compared with the periodic time of the modulating signal the voltage that appears across R_1 will include the following components: (*a*) the

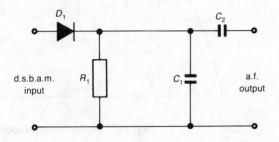

Fig. 1.6 Diode detector.

modulating signal, (*b*) a d.c. voltage that is directly proportional to the carrier amplitude, and (*c*) a number of high-frequency signals. The d.c. voltage is blocked by capacitor C_2 and the components (*c*) are filtered off.

Analysis

When the d.s.b.a.m. wave $[V_c + V_m(t)] \sin \omega_c t$ is rectified the output voltage consists of the envelope of half-sine waves at the carrier frequency. The Fourier series for a half-wave rectified sine wave is

$$ v = \frac{1}{\pi} + \frac{1}{2} \sin \omega_c t - \frac{2}{\pi} \left(\frac{\cos 2\omega_c t}{3} + \frac{\cos 4\omega_c t}{15} + \dots \right). $$

(1.18)

The output voltage of the detector is obtained by multiplying this series by $[V_c + V_m(t)]$. The product contains the terms $V_c/\pi + V_m(t)/\pi + \dots$; the first term is a d.c. voltage and the second term is the wanted detected audio signal. The *detection efficiency* η is the ratio

$$ \frac{\text{detected output voltage}}{\text{peak input voltage}} $$

expressed as a percentage. Figure 1.7 shows two consecutive half-cycles of the input voltage. If the input voltage is $V_c \cos \omega_c t$ the diode will conduct when $V_c \cos \omega_c t \geq V_{out}$; then $\omega_c t = \theta = \cos^{-1} (V_{out}/V_c)$ or $\cos \theta = V_{out}/V_c$. During the period 2θ the current that flows is

$$ i = \frac{V_c \cos \omega_c t - V_{out}}{r}, $$

where r is the forward resistance of the diode. The average value of the diode current is

$$ I_{DC} = \frac{1}{2\pi} \int_{-\pi}^{\pi} i \, d\omega_c t = \frac{1}{2\pi} \int_{-\theta}^{\theta} i \, d\omega_c t $$

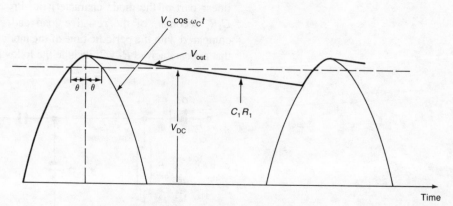

Fig. 1.7 Action of the diode detector.

$$= \frac{1}{\pi} \int_0^\theta i \, d\omega_c t$$

$$= \frac{1}{\pi r}(V_c \sin \theta - V_{out}\theta) = \frac{V_c}{\pi r}(\sin \theta - \theta \cos \theta).$$

Hence, the output voltage

$$V_{out} = I_{DC}R_1 = \frac{V_c R_1}{\pi r}(\sin \theta - \theta \cos \theta)$$

and

$$\eta = \frac{V_{out}}{V_c} = \cos \theta = \frac{R_1}{\pi r}(\sin \theta - \theta \cos \theta).$$

Then

$$\frac{\pi r}{R_1} = \tan \theta - \theta \simeq \theta + \frac{\theta^3}{3} - \theta = \frac{\theta^3}{3} \text{ or } \theta = \left(\frac{3\pi r}{R_1}\right)^{1/3}$$

and

$$\eta = \cos \left(\frac{3\pi r}{R_1}\right)^{1/3} \times 100\%$$

$$\simeq \left[1 - \frac{1}{2}\left(\frac{3\pi r}{R_1}\right)^{2/3}\right] \times 100\%. \tag{1.19}$$

For a low-voltage input signal the forward resistance of the diode is not constant and so the diode detector efficiency will vary. For voltages in excess of about 1 V the efficiency is constant at (usually) 90% or more.

The input resistance of a diode detector is related to its detection efficiency. The diode conducts only when the input voltage is at, or near, its peak positive value. Hence the average input power is

$$P_{IN} = V_c I_{DC} = \frac{\eta V_c^2}{R_1} = \left(\frac{V_c}{\sqrt{2}}\right)^2 \cdot \frac{1}{R_{IN}}.$$

Therefore,

$$R_{IN} = \frac{R_1}{2\eta}. \tag{1.20}$$

In modern radio receivers the reactive part of the detector low-pass filter is connected via a buffer amplifier so that the load seen by the diode detector is always resistive.

Example 1.7

Calculate the input resistance of a diode detector that has a load resistance of 8 kΩ and a diode forward resistance of 100 Ω.

Solution
From equation (1.19)

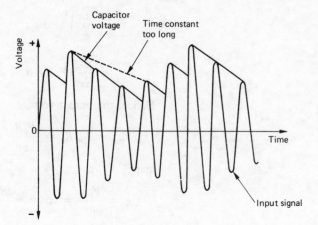

Fig. 1.8 Output voltage of a diode detector, showing the effect of an incorrect time constant.

$$\eta = \left[1 - \frac{1}{2} \left(\frac{3\pi \times 100}{8 \times 10^3} \right)^{2/3} \right] \times 100\% = 88\%.$$

From equation (1.20)

$$R_{IN} = \frac{8000}{2 \times 0.88} = 4545 \ \Omega. \quad (Ans.)$$

Clipping Caused by Incorrect Time Constant

The diode detector shunt capacitor C_1 must be able to discharge rapidly enough for the voltage across it to follow the modulation envelope. This is most difficult when the envelope is decreasing in amplitude, see Fig. 1.8. If the time constant C_1R_1 is too long, relative to the periodic time of the modulating signal, the capacitor voltage will not be able to follow the troughs of the modulation envelope; this is shown by the dotted line in Fig. 1.8.

As the capacitor C_1 discharges through R_1 its voltage is $v = V_c e^{-t/C_1R_1}$, where V_c is its initial, peak value which is equal to the peak input voltage (minus the diode voltage drop). The rate at which v falls is given by $-dv/dt = v/C_1R_1$. For v to follow the modulation envelope without distortion

$$\frac{v}{C_1R_1} \geq \frac{d}{dt} [V_c(1 + m \sin \omega_m t)] \geq -m\omega_m V_c \cos \omega_m t \quad V/s.$$

Thus

$$V_c(1 + m \sin \omega_m t) \geq -m\omega_m V_c \cos \omega_m t$$

or $- \dfrac{1}{C_1R_1} \geq \dfrac{m\omega_m V_c \cos \omega_m t}{V_c(1 + m \sin \omega_m t)}.$ \hfill (1.21)

This inequality is most difficult to satisfy when its right-hand side has its maximum negative value. Differentiating and equating to zero, gives $\sin \omega_m t = -m$. Therefore, $\cos \omega_m t = -\sqrt{(1 - m^2)}$, and substituting into inequality (1.21), gives

$$\frac{-1}{C_1 R_1} \geq \frac{-m\omega_m \sqrt{(1 - m^2)}}{1 - m^2} = \frac{-m\omega_m}{\sqrt{(1 - m^2)}}$$

and $C_1 R_1 \leq \dfrac{\sqrt{(1 - m^2)}}{m\omega_m}$. (1.22)

Example 1.8

A diode detector is to produce an audio output signal in the frequency band 100 to 4500 Hz. If the maximum modulation depth is 50% and the load resistor is 10 kΩ calculate the maximum possible value for the shunt capacitor.

Solution
From equation (1.22),

$$C_{1(max)} = \frac{\sqrt{(1 - 0.25^2)}}{0.5 \times 2\pi \times 4500 \times 10^4} = 6.85 \text{ nF.} \quad (Ans.)$$

The diode detector has the advantage of simplicity but the disadvantages that: (a) at least 0.5 V is required for the diode to conduct and this means that a high i.f. gain is necessary; (b) if the input signal-to-noise ratio is low the output signal-to-noise ratio will fall more rapidly than the input signal-to-noise ratio, this is known as the threshold effect; and (c) it generates energy at harmonics of the intermediate frequency of the radio receiver.

Non-linear Detection

If a d.s.b.a.m. signal is applied to a non-linear device detection will occur because of the v^2 term. The device may well be the diode in an envelope detector when the input voltage is small. If the input voltage is $V_c(1 + m \sin \omega_m t) \sin \omega_c t$ the v^2 term will give $bV_c^2(1 + m \sin \omega_m t)^2 \sin^2 \omega_c t$. Expanding this gives the term $bV_c^2 m \sin \omega_m t$ which is, of course, the wanted modulating signal. There are a number of other, unwanted, components at other frequencies also present and the most troublesome of these arises from the terms $bV_c^2/2 \; m^2 \sin^2 \omega_m t$. Expanding this gives $-bV_c^2/4 \; m^2 \cos 2\omega_m t$ which results in second-harmonic distortion and possible intermodulation.

The second-harmonic distortion is $\dfrac{bV_c^2 m^2/4}{bV_c^2 m}$ or $0.25 m\%$. If, for example, $m = 30\%$, the percentage second-harmonic distortion will be 7.5%. This second-harmonic component and the other, higher-frequency components will probably be removed by a low-pass filter. More serious are the intermodulation products that are generated.

Suppose that $V_m(t) = V_1 \sin \omega_1 t + V_2 \sin \omega_2 t$, then the squared term contains the component $bV_1V_2 \sin \omega_1 t \sin \omega_2 t$ which, upon expansion, shows the presence of components at frequencies $f_1 \pm f_2$. These intermodulation products will fall within the bandwidth occupied by the modulating signal and so they cannot be filtered out.

Coherent, Synchronous or Product Detector

A product detector multiplies together the d.s.b.a.m. wave to be demodulated and the unmodulated carrier; it is therefore necessary that the original carrier, at the correct frequency, is available at the receiver. The output voltage of the detector is $v_{out} = V_c \sin \omega_c t \times V_c(1 + m \sin \omega_m t) \sin \omega_c t$ or $V_c^2/2(1 - \cos 2\omega_c t)(1 + m \sin \omega_m t)$ and this contains the term $V_c V_m/2 \sin \omega_m t$. This is the wanted modulating signal so that detection has been achieved. The block diagram of a product detector is given by Fig. 1.9(a). The carrier component must be extracted from the incoming d.s.b.a.m. signal and one method of doing this is shown by Fig. 1.9(b). The d.s.b.a.m. signal is hard-limited to produce a square wave of frequency f_c. The two signals are applied to an analogue multiplier whose output contains the detected signal. Another carrier extraction method employs a phase-locked loop (see *Electronics IV*), see Fig. 1.10. The phase detector has inputs of $V_c(1 + m \sin \omega_m t) \sin \omega_c t$ and the voltage generated by the voltage-controlled oscillator (v.c.o.) and it generates an output voltage that is proportional to the phase difference between the two voltages. The output of the phase detector is fed, via a low-pass filter, to the control terminal of the v.c.o., causing it to change frequency in the direction which minimizes the error. Once *lock* has been established the v.c.o. frequency will be equal to the carrier frequency and this is also the output voltage of the circuit.

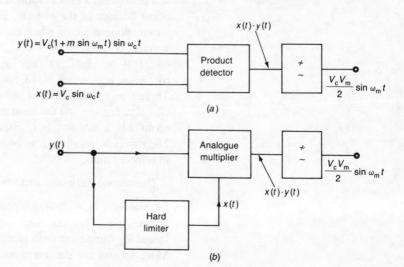

Fig. 1.9 (*a*) Product detector.
(*b*) Use of a hard limiter to obtain the carrier component.

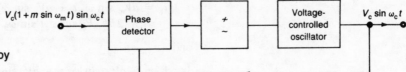

Fig. 1.10 Carrier extraction by phase-locked loop.

The product detector is now in widespread use since it is particularly convenient for implementation in an integrated circuit. Usually, the product detector is within the same IC package as other radio receiver circuitry. The product detector offers the advantage that the input signal may be very small since its lower limit is set only by the wanted signal-to-noise ratio. This will allow the i.f. gain to be up to 60 dB less than if a diode detector is used. The disadvantages are (*a*) the circuit will demodulate input noise if there is no signal and so a squelch circuit is really necessary, and (*b*) the d.c. output voltage is small and so a d.c. amplifier is needed to produce the a.g.c. voltage; this is often on-chip, if not an op-amp or an a.g.c. generator chip can be used.

Double-sideband Suppressed-carrier Amplitude Modulation

Most of the power carried by a d.s.b.a.m. wave is developed by the carrier frequency component and hence the transmission efficiency is low. Since the carrier conveys zero information it is not necessary that it be transmitted and it can be suppressed at the modulation stage if a *balanced modulator* is used. The basic principle of a balanced modulator is illustrated by Fig. 1.11. Each non-linear device, which may be a diode or a transistor, has a current-voltage characteristic given by $i = i_0 + av + bv^2 + \dots$. A FET is the best device for this purpose since its mutual characteristic very nearly obeys square-law. Assuming identical square-law devices, the voltage applied to n.l.d. A is $v_A = V_m \sin \omega_m t + V_c \sin \omega_c t$ and the voltage applied to n.l.d. B is $v_B = -V_m \sin \omega_m t + V_c \sin \omega_c t$. Therefore

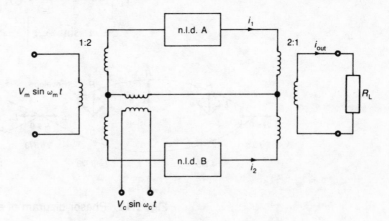

Fig. 1.11 Balanced modulator.

$$i_1 = aV_m \sin \omega_m t + aV_c \sin \omega_c t + bV_m^2 \sin^2 \omega_m t$$
$$+ 2bV_m V_c \sin \omega_c t \sin \omega_m t + bV_c^2 \sin^2 \omega_c t$$

and

$$i_2 = -aV_m \sin \omega_m t + aV_c \sin \omega_c t + bV_m^2 \sin^2 \omega_m t$$
$$- 2bV_m V_c \sin \omega_c t \sin \omega_m t + bV_c^2 \sin^2 \omega_c t.$$

The output current i_{out} of the modulator is proportional to the difference between i_1 and i_2 and so it is

$$i_{out} = 2(aV_m \sin \omega_m t + 2bV_m V_c \sin \omega_c t \sin \omega_m t).$$

The term at the modulating signal frequency can be removed by a filter, provided $\omega_c \gg \omega_m$, to leave the wanted d.s.b.s.c. signal. A transconductance multiplier IC can also be used as a balanced modulator; the carrier component voltage should be large enough to switch the transistors and the modulating signal input should be capacitor-coupled to remove any d.c. component. A number of IC balanced modulators are also available such as the Motorola 1596 and the Plessey SL 6401.

Phasor Diagram

The phasor diagram of a d.s.b.s.c. amplitude-modulated wave is shown in Fig. 1.12. Clearly, the envelope of the modulated wave is not sinusoidal, indicating that distortion has occurred. This means that the signal cannot be demodulated at the receiver unless the carrier component is re-inserted with *both* the correct frequency *and* phase. If there is an error θ in the phase of the re-inserted carrier the instantaneous voltage of the wave will be

$$v = V_c \sin (\omega_c t + \theta) + mV_c \sin \omega_c t \sin \omega_m t$$
$$= V_c[\cos \omega_c t \sin \theta + \sin \omega_c t(\cos \theta + m \sin \omega_m t)].$$

The envelope of this is $V_c\sqrt{[\sin^2 \theta + (\cos \theta + m \sin \omega_m t)^2]}$ or

$$V_c \left[\frac{1 - \cos 2\theta}{2} + \cos^2 \theta + 2 \cos \theta \, m \sin \omega_m t \right.$$
$$\left. + m^2 \sin^2 \omega_m t \right]^{1/2}$$

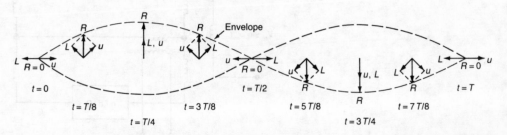

Fig. 1.12 Phasor diagram of a d.s.b.s.c. amplitude-modulated wave.

$$= V_c \left[\frac{1}{2} - \frac{\cos 2\theta}{2} + \frac{1}{2} + \frac{\cos 2\theta}{2} + 2m \cos \theta \sin \omega_m t \right.$$

$$\left. + \frac{m^2}{2} - \frac{m^2}{2} \cos 2\omega_m t \right]^{1/2}$$

$$= V_c \left[1 + \frac{m^2}{2} + 2m \cos \theta \sin \omega_m t - \frac{m^2}{2} \cos 2\omega_m t \right]^{1/2}$$

$$= V_c \sqrt{\left(1 + \frac{m^2}{2} \right)} \left[1 + \frac{4m \cos \theta \sin \omega_m t}{2 + m^2} \right.$$

$$\left. - \frac{m^2 \cos 2\omega_m t}{(2 + m^2)} + \dots \right]. \qquad (1.23)$$

Equation (1.23) shows that the envelope of the reconstructed d.s.b.s.c. a.m. waveform varies in a manner that is a function of *both* the modulating signal *and* the phase error. If $\theta = 90°$ $\cos \theta = 0$ and there will be no variation of the envelope at the signal frequency. What amplitude modulation there is occurs at twice the signal frequency and there is considerable phase modulation as well.

Example 1.9

A 20 V carrier is amplitude modulated to a depth of 60%. The carrier is then removed and after a phase shift of 90° is re-inserted. Calculate the resulting peak phase deviation.

Solution
The two side-frequency phasors are $(0.6 \times 20)/2 = 6$ V. The peak phase deviation is, see Fig. 1.13,

$$\phi_{max} = \tan^{-1} 12/20 = 31°. \quad (Ans.)$$

The requirement for the re-inserted carrier to be correct in both frequency and phase is not simple to satisfy and requires the use of complex circuitry. The d.s.b.s.c. system is not used for ordinary radiocommunication purposes but it is employed for (*a*) the transmission of colour information in television broadcasting, and (*b*) the transmission of stereo information in v.h.f. sound broadcasting.

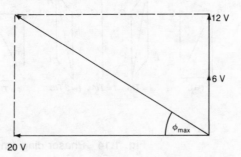

Fig. 1.13

Single-sideband Suppressed-carrier Amplitude Modulation

The information carried by an amplitude-modulated wave is contained in *both* sidebands and so it is not necessary for both sidebands to be transmitted. Either sideband can be suppressed at the transmitter without any loss of data. Figure 1.14 shows the phasor diagram of an s.s.b.s.c. a.m. signal with its carrier component re-inserted with the correct phase. The envelope can be seen to be sinusoidal. The maximum phase error is 90° and if the phasor diagram is redrawn with this carrier re-insert error it will be found that the envelope is still sinusoidal. This means that the phase of the re-inserted carrier is not important for a speech system.

Two methods of suppressing the unwanted sideband are used: either the output of the balanced modulator can be passed through a filter of the appropriate bandwidth, or the phasing technique shown in Fig. 1.15 can be used. The use of a filter to remove the unwanted sideband becomes more difficult at high frequencies because the frequency gap between the sidebands becomes a small percentage of the filter's centre frequency. For this reason modulation is often carried out at a low frequency, then the wanted sideband is shifted to the desired position in the frequency spectrum.

Referring to Fig. 1.15, the two balanced modulators have both the modulating and carrier voltages applied to them, but the upper modulator has both its inputs phase shifted by 90° before they are applied. The input signals to the upper modulator are $V_m \sin (\omega_m t + 90°)$ and $V_c \sin (\omega_c t + 90°)$ so that its output current contains components

$$\cos [(\omega_c t + 90°) - (\omega_m t + 90°)] - \cos [(\omega_c t + 90°) + (\omega_m t + 90°)]$$
$$= \cos (\omega_c t - \omega_m t) - \cos (\omega_c t + \omega_m t + 180°).$$

The inputs to the lower modulator are $V_m \sin \omega_m t$ and $V_c \sin \omega_c t$ and so its output current includes components $\cos (\omega_c t - \omega_m t) - \cos (\omega_c t + \omega_m t)$. The output signals of the two modulators are added together and, since the upper side-frequency components are in antiphase with one another they cancel, give an output of $2 \cos (\omega_c - \omega_m)t$. If it is required to transmit the upper sideband instead

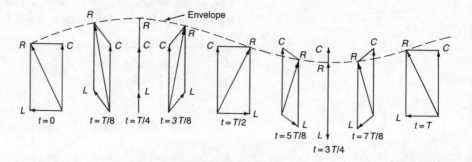

Fig. 1.14 Phasor diagram of an s.s.b.s.c. a.m. signal.

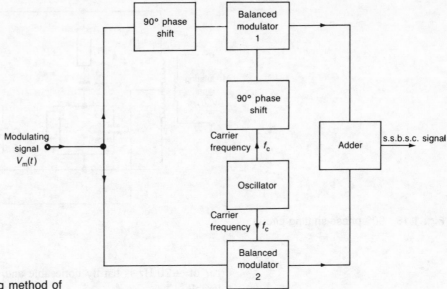

Fig. 1.15 The phasing method of producing an s.s.b.s.c. a.m. signal.

of the lower sideband *either* one of the 90° phase-shifting circuits must be moved to the lower part of the circuit. The phasing method of suppressing one sideband has the advantage that it is easy to switch from transmitting one sideband to transmitting the other. One circuit which can be used as the audio phase-shifting circuit is shown by Fig. 1.16; here $\omega_1 = 1/C_1R_1$ and $\omega_2 = 1/C_2R_2$ specify the audio bandwidth shifted.

Demodulation of s.s.b.s.c. Signals

An s.s.b.s.c. wave can be demodulated using either a product, or a switching, detector. Suppose that the lower sideband is received. The output of a product detector is proportional to the product of the lower sideband signal and a locally generated carrier. Thus,

$$v_{out} = V \sin (\omega_c - \omega_m)t \; V_c \sin \omega_c t$$
$$= VV_c(\sin \omega_c t \cos \omega_m t - \sin \omega_m t \cos \omega_c t) \sin \omega_c t.$$

This includes the term $VV_c/2 \cos \omega_m t$ which is the demodulated message signal and it can be extracted using a low-pass filter. If there is a frequency error $\delta\omega_c$ in the re-inserted carrier the wanted output of the detector will be $VV_c/2 \cos (\omega_m t + \delta\omega_c t)$. The frequency error that can be tolerated depends upon the type of signal involved. Since every frequency contained in the modulating signal is shifted by the same amount δf_c the harmonic relationship between the components is lost. The recommendation of the CCITT is that the error should not be in excess of ± 2 Hz. However, it is found that for speech an

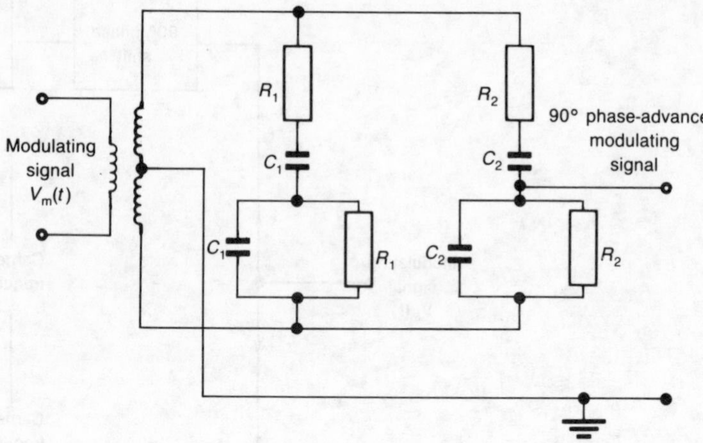

Fig. 1.16 90° phase-shifting circuit.

error of ± 20 Hz is hardly noticeable and as much as ± 50 Hz is tolerable.

If the re-inserted carrier is of the correct frequency but there is a phase error θ the demodulated signal will contain a term $VV_c/2 \cos (\omega_m t - \theta)$. For speech and music this error is of little consequence but it will matter for any system where the signal waveshape is of importance.

Single-sideband Compared with Double-sideband Amplitude Modulation

Single-sideband operation of a radio system offers a number of advantages over d.s.b. operation.

(a) The bandwidth required for an s.s.b.s.c. system is only one-half of the bandwidth that must be allocated to a d.s.b.a.m. system. The reduction in bandwidth allows a greater number of channels to be accommodated within a given bandwidth.

(b) The signal-to-noise ratio at the output of an s.s.b.s.c. system is higher than the output signal-to-noise ratio of a d.s.b. system transmitting the same power. If the carrier is sinusoidally modulated the d.s.b. system will have peak side-frequency voltages of $mV_c/2$ and a peak envelope voltage of $V_c(1 + m)$ volts. The s.s.b.s.c. system will have a side-frequency voltage of $V_c(1 + m)$ volts; this is $V_c(1 + m)/(mV_c/2)$ or $2(1 + m)/m$ times as great as each d.s.b. side-frequency voltage. If the transmission path is free from distortion the two d.s.b. side-frequency voltages add algebraically in the detection process. Hence the s.s.b.s.c. side-frequency voltage is $(1 + m)/m$ times as great as the d.s.b. sum voltage. Quoted in decibels, this is $20 \log_{10}[(1 + m)/m]$ dB. In addition, the output noise power will be reduced by 3 dB because the bandwidth has been halved. Therefore, the increase in the output signal-to-noise ratio given by s.s.b.s.c. operation is

$$\text{signal-to-noise ratio increase} = 3 + 20 \log_{10}\left(\frac{1 + m}{m}\right) \text{ dB.}$$

$$(1.24)$$

The maximum possible value for the modulation factor m is unity and this gives the signal-to-noise ratio improvement as 9 dB. For any smaller value of m an even larger advantage is obtained.

(c) An s.s.b.s.c. transmitter is more efficient than a d.s.b. transmitter.

(d) Selective fading of d.s.b. radio waves may cause considerable distortion when the carrier component fades relative to the side-frequencies. This effect does not occur in an s.s.b.s.c. system because the received signal is demodulated against a re-inserted carrier of constant amplitude.

Example 1.10

A d.s.b.a.m. system radiates a carrier power of 30 kW. The system is changed to operate using s.s.b.s.c. and a radiated power of 10 kW. Calculate the change in the output signal-to-noise ratio of the system if the maximum depth of modulation is 75%.

Solution

From equation (1.24), the increase in the output signal-to-noise ratio is $3 + 20 \log_{10}(1.75/0.75) = 10.36$ dB. The decrease in the transmitted power is $10 \log_{10}(30/10) = 4.77$ dB. Therefore, increase in output signal-to-noise ratio

$$= 10.36 - 4.77 = 5.59 \text{ dB.} \quad (Ans.)$$

Peak Envelope Power

The output power of an s.s.b.s.c. radio transmitter is usually specified in terms of its *peak envelope power* (p.e.p.). The p.e.p. is the power developed by the peak value of the transmitted sideband and any pilot carrier. When there is no pilot carrier, or it is of very small amplitude, the term *peak sideband power* (p.s.p.) is often employed instead.

2 Angle Modulation

When a sinusoidal carrier wave $V_c \sin(\omega_c t + \theta)$ is *angle* modulated the amplitude V_c of the carrier is maintained at a constant value and its instantaneous phase $\omega_c t + \theta$ is varied by the modulating signal. There are two possibilities: either the frequency $\omega_c/2\pi$ of the carrier wave, or its phase θ, can be made to vary in direct proportion to the instantaneous voltage of the modulating signal. The differences between *frequency modulation* and *phase modulation* are not obvious, since a change in frequency must inherently involve a change in phase, and they are listed in Table 2.2.

Frequency modulation is much more commonly used than phase modulation and it possesses a number of advantages over amplitude modulation, particularly if a wide bandwidth can be made available. Frequency modulation is employed for sound broadcasting in the v.h.f. band, for the sound signal of u.h.f. television broadcast transmissions, for some land, sea and air mobile systems, and for u.h.f./s.h.f. multi-channel telephony systems including those routed via communications satellites. Phase modulation is not often used in analogue radio communication and it finds its main application in the field of digital radio communication. It is, however, often employed as a stage in the generation of a frequency-modulated wave.

Frequency Modulation

Frequency modulation of a carrier wave occurs when the instantaneous deviation from the unmodulated carrier frequency is directly proportional to the instantaneous amplitude of the modulating signal voltage. The maximum amount by which the carrier frequency can be deviated is known as the *frequency deviation*. Frequency deviation has no inherent limit and for any particular f.m. system a maximum permissible deviation must be specified. This maximum is known as the *rated system deviation* f_d. Once the rated system deviation has been determined it sets the maximum modulating voltage that can be applied to the frequency modulator. Most of the time the modulating voltage will be smaller than this maximum value and then the frequency deviation is kf_d, where

$$k = \frac{\text{modulating signal voltage}}{\text{maximum allowable modulating signal voltage}} \quad (2.1)$$

Clearly, k can have any value between zero and unity.

Instantaneous Voltage

If the modulating signal is $V_m(t)$ the instantaneous carrier frequency f_i will be

$$f_i = f_c + kf_d V_m(t). \tag{2.2}$$

The instantaneous angular velocity ω_i of the modulated wave is

$$\omega_i = 2\pi f_i = d\theta/dt = 2\pi f_c + 2\pi kf_d V_m(t),$$

and

$$\theta = \int_0^t \omega_i dt = \int_0^t 2\pi f_c dt + \int_0^t 2\pi kf_d V_m(t) \, dt, \text{ or}$$

$$\theta = \omega_c t + 2\pi kf_d \int_0^t V_m(t) \, dt. \tag{2.3}$$

The instantaneous voltage v_c of the frequency-modulated carrier wave is $v_c = V_c \sin\theta$ or

$$v_c = V_c \sin\left[\omega_c t + 2\pi kf_d \int_0^t V_m(t) \, dt\right]. \tag{2.4}$$

The peak phase deviation of the carrier depends upon the integral with respect to time of the modulating signal voltage.

If the modulating signal voltage is of sinusoidal waveform, i.e. $V_m(t) = V_m \cos \omega_m t$, then equations (2.2) and (2.3) become

$$f_i = f_c + kf_d \cos \omega_m t \tag{2.5}$$

and $\theta = \omega_c t + \dfrac{kf_d}{f_m} \sin \omega_m t,$ $\tag{2.6}$

respectively, and the instantaneous voltage of the wave is

$$v_c = V_c \sin\left[\omega_c t + \frac{kf_d}{f_m} \sin \omega_m t\right]. \tag{2.7}$$

Modulation Index

The peak phase deviation of the carrier is equal to kf_d/f_m and this factor is usually known as the *modulation index* m_f. The modulation index is equal to the ratio of the frequency deviation to the modulating frequency. Very often the expression for the instantaneous voltage of a sinusoidally modulated f.m. wave is written in terms of the modulation index, thus

$$v_c = V_c \sin(\omega_c t + m_f \sin \omega_m t). \tag{2.8}$$

Deviation Ratio

When an f.m. system is designed the maximum permissible values for both the frequency deviation of the carrier and the modulating signal frequency must be used. Then the modulation index is known as the *deviation ratio D*.

$$D = f_d/f_{m(max)}. \tag{2.9}$$

The deviation ratio of a particular f.m. system is fixed, whereas the modulation index varies continuously with change in the modulating signal voltage and/or frequency. In the BBC v.h.f. sound broadcast system $f_d = 75$ kHz and $f_{m(max)} = 15$ kHz so that $D = 5$.

Example 2.1

When a 2 kHz sinusoidal signal is applied to a frequency modulator the 90 MHz carrier is deviated by ± 16 kHz. (*a*) Calculate the phase deviation of the carrier. (*b*) Calculate the new frequency and phase deviations if the modulating signal has both its voltage and its frequency doubled.

Solution

(*a*) Phase deviation = $(16 \times 10^3)/(2 \times 10^3) = 8$ rad.   (*Ans.*)
(*b*) Phase deviation = $(32 \times 10^3)/(4 \times 10^3) = 8$ rad.   (*Ans.*)
    Frequency deviation = 32 kHz.   (*Ans.*)

Frequency Spectrum of a Frequency-modulated Wave

The frequency spectrum of a frequency-modulated wave is much more complicated than that of an a.m. wave. For small values of modulation index, say $m_f \leq 0.25$, the f.m. wave consists of a carrier component f_c and two side-frequencies $f_c \pm f_m$, just like a d.s.b.a.m. waveform. If, however, the modulation index is increased, second-order side-frequencies $f_c \pm 2f_m$ will also appear. Further increase in the value of m_f leads to the appearance of more, and more, higher-orders of side-frequencies, and the frequency spectrum rapidly becomes complex. Equation (2.8) can be rewritten (using the identity $\sin (A + B) = \sin A \cos B + \sin B \cos A$) in the form

$$\begin{aligned} v_c = {} & V_c \sin \omega_c t \cos (m_f \sin \omega_m t) \\ & + V_c \cos \omega_c t \sin (m_f \sin \omega_m t). \end{aligned} \tag{2.10}$$

If the value of m_f is less than unity equation (2.10) can be expanded using the series forms of $\sin \theta$ and $\cos \theta$, but if m_f is larger than unity expansion requires the use of Bessel functions.

Small values of m_f

The series forms of $\sin \theta$ and $\cos \theta$ are

$$\sin \theta = \theta - \frac{\theta^3}{3!} + \frac{\theta^5}{5!} - \dots,$$

and

$$\cos \theta = 1 - \frac{\theta^2}{2!} + \frac{\theta^4}{4!} \dots.$$

If m_f is very small, say less than 0.25, m_f^2 and all higher powers of m_f will be negligibly small, and then

$$\cos (m_f \sin \omega_m t) \simeq 1$$

and $\sin (m_f \sin \omega_m t) \simeq m_f \sin \omega_m t$.

Equation (2.10) can then be written as

$$
\begin{aligned}
v_c &= V_c \sin \omega_c t + V_c m_f \cos \omega_c t \sin \omega_m t \\
&= V_c \sin \omega_c t + m_f V_{c/2}[\sin (\omega_c + \omega_m)t \\
&\quad - \sin (\omega_c - \omega_m)t],
\end{aligned}
\tag{2.11}
$$

(using the identity $2 \cos A \sin B = \sin (A + B) - \sin (A - B)$.

This shows that the spectrum of a narrow-band frequency-modulated (n.b.f.m.) system consists of the carrier f_c and the lower and upper side-frequencies $f_c \pm f_m$. Comparing with equation (1.4) it can be seen that n.b.f.m. is equivalent to d.s.b.a.m. with the lower side-frequency phase-shifted by 180°.

$0.25 \le m_f \le 1$

If the modulation index m_f is greater than 0.25 and less than unity,

$$\cos (m_f \sin \omega_m t) \simeq 1 - \frac{m_f^2 \sin^2 \omega_m t}{2} \text{ and}$$

$$\sin (m_f \sin \omega_m t) \simeq m_f \sin \omega_m t.$$

This means that the instantaneous voltage of the f.m. wave is

$$
\begin{aligned}
v_c &= V_c \sin \omega_c t \left[\frac{1 - m_f^2 \sin^2 \omega_m t}{2} \right] \\
&\quad + m_f V_c \cos \omega_c t \sin \omega_m t \\
&= V_c \sin \omega_c t - \frac{m_f^2 V_c \sin \omega_c t}{4} \\
&\quad + \frac{m_f^2 V_c \sin \omega_c t \cos 2\omega_m t}{4} + m_f V_c \cos \omega_c t \sin \omega_m t.
\end{aligned}
$$

Hence (using the identity $2 \cos A \sin B = \sin (A + B) - \sin (A - B)$),

$$
\begin{aligned}
v_c &= V_c \sin \omega_c t \left[1 - \frac{m_f^2}{4} \right] + \frac{m_f V_c}{2} [\sin (\omega_c + \omega_m)t \\
&\quad - \sin (\omega_c - \omega_m)t] + \frac{m_f^2 V_c}{8} [\sin (\omega_c + 2\omega_m)t \\
&\quad + \sin (\omega_c - 2\omega_m)t].
\end{aligned}
\tag{2.12}
$$

Example 2.2

A 10 V, 100 MHz carrier wave is frequency modulated by a 2000 Hz tone when the frequency deviation is 1000 Hz. Calculate the amplitudes of the first-order, and of the second-order side-frequencies.

Solution

$m_f = kf_d/f_m = 0.5$. From equation (2.12),

$$\text{Carrier amplitude} = 10 \left(1 - \frac{0.5^2}{4}\right) = 9.375 \text{ V.} \quad (Ans.)$$

$$\text{First-order side-frequency amplitude} = \frac{0.5 \times 10}{2} = 2.5 \text{ V.}$$
$$(Ans.)$$

$$\text{Second-order side-frequency amplitude} = \frac{0.5^2 \times 10}{8} = 0.3125 \text{ V.}$$
$$(Ans.)$$

$m_f \geq 1$

When the modulation index m_f is equal to, or is greater than, unity expansion of equation (2.10) requires the use of Bessel functions. Using these shows the presence of a number of higher-order side-frequencies in the f.m. wave. This is shown by equation (2.13).

$$\begin{aligned}
v_c =\ & V_c J_0(m_f) \sin \omega_c t \\
& + V_c J_1(m_f) [\sin (\omega_c + \omega_m)t - \sin (\omega_c - \omega_m)t] \\
& + V_c J_2(m_f) [\sin (\omega_c + 2\omega_m)t + \sin (\omega_c - 2\omega_m)t] \\
& + V_c J_3(m_f) [\sin (\omega_c + 3\omega_m)t - \sin (\omega_c - 3\omega_m)t] \\
& + V_c J_4(m_f) [\sin (\omega_c + 4\omega_m)t + \sin (\omega_c - 4\omega_m)t] \\
& + \text{etc.,} \quad\quad\quad\quad\quad\quad\quad\quad\quad\quad\quad\quad\quad (2.13)
\end{aligned}$$

where $J_0(m_f)$, $J_1(m_f)$, $J_2(m_f)$, etc., are Bessel functions of order 0, 1, 2, etc., and give the amplitudes of the carrier and side-frequency components. Examination of equation (2.13) shows that the frequency spectrum contains the carrier component at frequency f_c, the odd-order side-frequencies $f_c \pm f_m$, $f_c \pm 3f_m$, $f_c \pm 5f_m$, etc., and even-order side-frequencies $f_c \pm 2f_m$, $f_c \pm 4f_m$, $f_c \pm 6f_m$, etc. All the odd-order lower side-frequencies have a *minus* sign in front of the lower side-frequency component, whilst all the even-order lower side-frequency components are preceded by a *plus* sign.

The amplitudes of the carrier, and of the various side-frequencies, vary with the modulation index but not in a simple manner. The carrier and each order of side-frequency reach a positive maximum value and then decrease and go cyclically through zero, reach a negative maximum value and then rise towards zero, and so on. Side-frequencies of the same order have the same amplitude so that the frequency spectrum is symmetrical about the carrier frequency. Table 2.1 gives the values of $J_0(m_f)$, $J_1(m_f)$, etc. for integer values of m_f from 1 to 12 and for side-frequencies up to the ninth order. The data is also shown graphically by Fig. 2.1. Table 2.2 shows the values of the modulation index, or the deviation ratio, at which the carrier, the first-, second- and third-order side-frequencies are zero.

Table 2.1

Component	$J_n(1)$	$J_n(2)$	$J_n(3)$	$J_n(4)$	$J_n(5)$	$J_n(6)$	$J_n(7)$	$J_n(8)$	$J_n(9)$	$J_n(10)$	$J_n(11)$	$J_n(12)$
Carrier $[J_0(m_f)]$	0.7652	0.2239	−0.2601	−0.3971	−0.1776	0.1506	0.3001	0.1717	−0.0903	−0.2459	−0.1712	0.0477
1st order	0.4401	0.5767	0.3391	−0.0660	−0.3276	−0.2767	−0.0047	0.2346	0.2453	0.0435	−0.1768	−0.2234
2nd order	0.1149	0.3528	0.4861	0.3641	0.0466	−0.2429	−0.3014	−0.1130	0.1448	0.2546	0.1390	−0.0849
3rd order	0.0196	0.1289	0.3091	0.4302	0.3648	0.1148	−0.1676	−0.2911	−0.1809	0.0584	0.2273	0.1951
4th order	—	0.0340	0.1320	0.2811	0.3912	0.3576	0.1578	−0.1054	−0.2655	−0.2196	−0.0150	0.1825
5th order	—	—	0.0430	0.1321	0.2611	0.3621	0.3479	0.1858	−0.0550	−0.2341	−0.2383	−0.0735
6th order	—	—	0.0114	0.0491	0.1310	0.2458	0.3392	0.3376	0.2043	−0.0145	−0.2016	−0.2437
7th order	—	—	—	0.0152	0.0534	0.1296	0.2336	0.3206	0.3275	0.2167	0.0184	−0.1703
8th order	—	—	—	—	0.0184	0.0565	0.1280	0.2235	0.3051	0.3179	0.2250	0.0451
9th order	—	—	—	—	—	0.0212	0.0589	0.1263	0.2149	0.2919	0.3089	0.2304

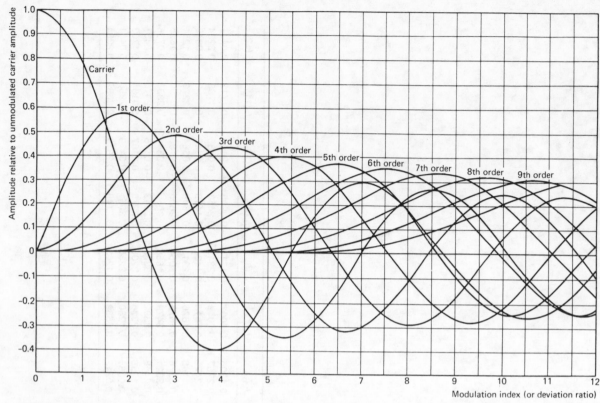

Fig. 2.1 Showing how the amplitudes of the various components of an f.m. wave vary with the modulation index.

Table 2.2

Modulation index m_f

Carrier	First-order	Second-order	Third-order
2.405	3.832	5.136	6.380
5.520	7.016	8.417	9.761
8.754	10.173	11.620	13.015
11.792	13.324	14.796	16.223

Figure 2.1 can be employed to determine the amplitudes, relative to the unmodulated carrier voltage, of each of the components of an f.m. wave. The amplitude of a component, for a particular value of m_f, or D, is obtained from the figure by projecting vertically upwards from the horizontal axis on to the required carrier, or nth-order side-frequency, curve and thence on to the vertical axis. Negative signs are (usually) ignored since only the magnitude of each component is (usually) required. The number of significant side-frequencies, i.e. those of amplitude ± 0.01 or greater, is equal to $(m_f + 1)$.

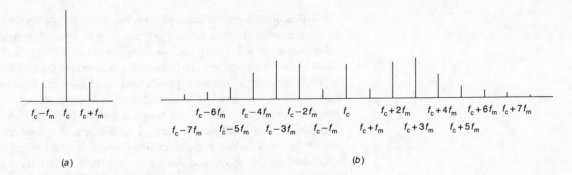

f_c-f_m f_c f_c+f_m

(a)

f_c-6f_m f_c-4f_m f_c-2f_m f_c f_c+2f_m f_c+4f_m f_c+6f_m f_c+7f_m

f_c-7f_m f_c-5f_m f_c-3f_m f_c-f_m f_c+f_m f_c+3f_m f_c+5f_m

(b)

Fig. 2.2 Spectrum diagrams for f.m. waves (a) $m_f = 0.5$, (b) $m_f = 4$.

Example 2.3

Plot the frequency spectrum diagrams of a frequency-modulated wave having a deviation ratio of (a) 0.5 and (b) 4.

Solution

The required spectrum diagrams are shown in Figs 2.2(a) and (b) respectively. The amplitudes of the components of Fig. 2.2(a) were calculated in Example 2.2 and should be compared.

The Phasor Diagram of an f.m. Wave

The phasor diagram of an n.b.f.m. wave, that has first-order side-frequencies only, has its lower and upper side-frequency phasors positioned symmetrically about a quadrature carrier phasor. This is shown by Fig. 2.3 for time intervals of $T/8$, where T is the periodic time of the modulating signal. As with the phasor diagram of the a.m. wave it has been assumed that the carrier phasor is stationary in the vertical direction so that the lower and upper side-frequency phasors rotate in opposite directions with angular velocity $\pm\omega_m$. The phase of the resultant, relative to the unmodulated carrier phasor is varied in a linear manner either side of the carrier. The amplitude of the resultant phasor is *not* constant, which means that some amplitude modulation is present.

The phasor diagram of an f.m. wave having both first-order and second-order side-frequencies is more complex. The second-order

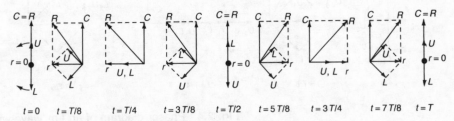

$t=0$ $t=T/8$ $t=T/4$ $t=3T/8$ $t=T/2$ $t=5T/8$ $t=3T/4$ $t=7T/8$ $t=T$

Fig. 2.3 Phasor diagram of an n.b.f.m. wave.

side-frequencies have the same phase relationship to one another as do the a.m. side-frequencies and they rotate with twice the angular velocity, i.e. at $\pm 2\omega_m$. Figure 2.4 shows the phasor diagram. The effect of the second-order side-frequencies is to reduce the variations in the length of the resultant phasor and so reduce the amplitude modulation of the carrier. The second-order side-frequencies do not affect the phase deviation of the resultant phasor from the carrier.

In general, odd-order side-frequencies contribute to the phase deviation of the carrier whilst even-order side-frequencies tend to keep the amplitude of the modulated waveform constant. If *all* the side-frequencies of an f.m. wave are included, the phasor diagram will have a constant-length resultant phasor R which deviates either side of the carrier phasor; this is shown by Fig. 2.5. The maximum phase deviation ϕ_{max} occurs when all the odd-order side-frequency phasors are instantaneously in phase with one another and in phase quadrature with the carrier phasor.

Complex Modulating Signal

When a carrier is frequency modulated by a complex signal the resulting frequency spectrum is *not* the sum of the spectra produced by each component in the complex signal acting alone. Consider a carrier frequency modulated by the two-tone signal $V_m(t) = V_1 \sin \omega_1 t + V_2 \sin \omega_2 t$. Each component acting on its own will give a certain modulation index m and will produce a carrier and various side-frequency components. The magnitudes of these components can be obtained from Table 2.1. If the modulation index due to $V_1 \sin \omega_1 t$ acting alone is m_{f1} then the frequency components in the f.m. wave are J_{01} and J_{m1}; similarly $V_2 \sin \omega_2 t$ will give a modulation index of m_{f2} and components J_{02} and J_{m2}. For example, suppose that $m_{f1} = 2$, then from Table 2.1 $J_{01} = 0.2239$ and $J_{m1} = 0.5767, 0.3528, 0.1289$ and 0.0340.

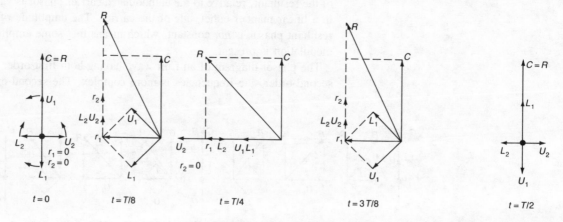

Fig. 2.4 Phasor diagram of an f.m. wave with both first-order and second-order side-frequencies.

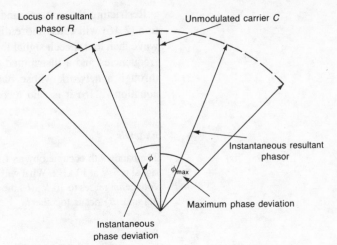

Fig. 2.5 Resulting phasor diagram of an f.m. wave when all side-frequencies are included.

When the carrier is modulated by the two-tone signal, the f.m. wave will contain components at the following frequencies:

(a) carrier frequency f_c of amplitude $J_{01} \times J_{02}$;
(b) side-frequencies $f_c \pm m_{f1}f_1$ of amplitude $J_{m1} \times J_{02}$;
(c) side-frequencies $f_c \pm m_{f2}f_2$ of amplitude $J_{m2} \times J_{01}$; and
(d) side-frequencies $f_c \pm m_{f1}f_1 \pm m_{f2}f_2$ of amplitude $J_{m1} \times J_{m2}$.

Clearly, even two-tone modulation will give rise to a large number of frequencies in the f.m. wave.

Bandwidth Required for an f.m. Wave

It is difficult to determine the exact bandwidth that is occupied by an f.m. wave because of the large number of side-frequencies it may contain. Usually, it is necessary to band-limit the wave and not transmit all of the higher-order side-frequencies. Between 98% and 99% of the signal power will be retained if $(m_f + 1)$ pairs of side-frequencies are transmitted. The minimum bandwidth required is hence

$$\text{bandwidth} = 2f_m(m_f + 1) \tag{2.14}$$

$$\text{or} \quad \text{bandwidth} = 2(kf_d + f_m). \tag{2.15}$$

An f.m. system will, of course, be designed to transmit the most demanding modulating signal without excessive distortion. This signal is the one that produces the rated system deviation and which contains components at up to the maximum frequency $f_{m(max)}$ to be transmitted. The minimum bandwidth that must be provided for the transmission of this signal is given by equation (2.16), i.e.

$$\text{system bandwidth} = 2(f_d + f_{m(max)}). \tag{2.16}$$

Restricting the system bandwidth to the value specified by equation (2.16) will cause more distortion to the sinusoidally modulated wave than to a speech signal that contains random voltages at random frequencies and a distributed spectrum. If an f.m. signal is passed through a network whose bandwidth is smaller than specified by equation (2.16) it is said to be *over-deviated*.

Example 2.4

The bandwidth occupied by an f.m. signal is 80 kHz when the modulating signal is 5 V at 10 kHz. What will be the occupied bandwidth if (a) the signal voltage increases to 10 V, (b) the frequency decreases to 5 kHz, or (c) both (a) and (b) occur together?

Solution
From equation (2.15), 80 kHz = $2(kf_d + 10)$ kHz or $kf_d = 30$ kHz.

(a) New $kf_d = 10/5 \times 30 = 60$ kHz; $B = 2(60 + 10) = 140$ kHz.
(*Ans.*)

(b) $B = 2(30 + 5) = 70$ kHz.   (*Ans.*)

(c) $B \doteq 2(60 + 5) = 130$ kHz.   (*Ans.*)

Phase Modulation

When a carrier is phase modulated the phase deviation of the carrier is directly proportional to the instantaneous amplitude of the modulating signal. Thus

$$\theta(t) = \omega_c t + k\Phi_d V_m(t), \tag{2.17}$$

where Φ_d is the peak phase deviation permitted in the system, known as the rated system deviation, and k (as for frequency modulation) is the ratio (modulating signal voltage)/(maximum permissible modulating signal voltage). The instantaneous voltage of a phase-modulated wave is

$$v_c = V_c \sin\left[\omega_c t + k\Phi V_m(t)\right] \tag{2.18}$$

or $$v_c = V_c \sin\left[\omega_c t + m_p V_m(t)\right], \tag{2.19}$$

where m_p is the modulation index and is equal to the peak phase deviation of the carrier.

The instantaneous angular velocity ω_i of the phase-modulated wave is the rate of change of its phase, so

$$\omega_i = \frac{d}{dt}\left[\omega_c t + k\Phi_d V_m(t)\right] = \omega_c + k\Phi_d \frac{dV_m(t)}{dt}$$

and the instantaneous frequency f_i is

$$f_i = f_c + \frac{k\Phi_d d V_m(t)}{2\pi dt}. \tag{2.20}$$

If the modulating signal is of sinusoidal waveform so that $V_m(t) = V_m \sin \omega_m t$, then

$$\theta = \omega_c t + k\Phi_d \sin \omega_m t \tag{2.21}$$

$$v_c = V_c \sin (\omega_c t + k\Phi_d \sin \omega_m t) \tag{2.22}$$

and

$$f_i = f_c + k\Phi_d f_m \cos \omega_m t. \tag{2.23}$$

The differences between frequency modulation and phase modulation can now be seen by comparing equations (2.5) and (2.6) with (2.21) and (2.23). The results are tabulated in Table 2.3.

Table 2.3

Modulation	Frequency deviation	Phase deviation
Frequency	Proportional to the voltage of the modulating signal	Proportional to the voltage and inversely proportional to the frequency of the modulating signal
Phase	Proportional to both the voltage and the frequency of the modulating signal	Proportional to the voltage of the modulating signal

Equation (2.22) is of the same form as equation (2.8), where $m_p = k\Phi_d$, and it can be similarly expanded to show the presence of various orders of side-frequencies. Although the two forms of angle modulation are similar, frequency modulation is much more common. The use of phase modulation is restricted to some mobile systems and to high-speed data circuits. Phase modulation is more difficult to detect since it requires an accurate reference phase; also it makes less efficient use of an available bandwidth than does frequency modulation, and so its signal-to-noise ratio is inferior.

Consider two systems, one frequency and the other phase modulated, having a maximum modulating frequency of 15 kHz and a modulation index of 5. The number of significant side-frequencies is then 6 so that a minimum bandwidth of $6 \times 15 \times 2 = 180$ kHz must be provided. If the modulating frequency is reduced, in turn, to 5, 3 and 1 kHz the modulation index will increase in the f.m. system but will remain constant in the p.m. system. Consequently, the number of significant side-frequencies, and hence the occupied bandwidth, will vary as shown by Table 2.4.

Table 2.4

| Modulating signal frequency (kHz) | Frequency modulation | | Phase modulation | |
	Modulation index	Occupied bandwidth (kHz)	Modulation index	Occupied bandwidth (kHz)
15	5	$6 \times 15 \times 2 = 180$	5	$6 \times 15 \times 2 = 180$
5	15	$16 \times 5 \times 2 = 160$	5	$6 \times 5 \times 2 = 60$
3	25	$26 \times 3 \times 2 = 156$	5	$6 \times 3 \times 2 = 36$
1	75	$76 \times 1 \times 2 = 152$	5	$6 \times 1 \times 2 = 12$

Signal-to-noise Ratio in an f.m. System

During the transmission of a frequency-modulated signal both noise and interference will be picked up and will both amplitude and phase modulate the signal. The first of these effects is easily removed by the use of a *limiter* in the f.m. receiver. Phase modulation of an f.m. signal, on the other hand, cannot be removed and it is only possible to minimize its effects upon the output signal-to-noise ratio of the system.

Consider the unmodulated carrier frequency and suppose that the interfering signal is also of sinusoidal waveform. If the voltages of the carrier and noise signals are V_c and V_n respectively, the phasor diagram is given by Fig. 2.6. In general, the noise voltage will not be at the same frequency as the unmodulated carrier and so its phasor will rotate about the top of the carrier phasor with an angular velocity ω_{diff}, where ω_{diff} is 2π times the difference in their frequencies. If the noise voltage is at a higher frequency than the carrier voltage the noise phasor will rotate in the anti-clockwise direction as shown.

The total voltage is

$$v = V_c \sin \omega_c t + V_n \sin (\omega_c + \omega_{\text{diff}})t$$
$$= V_c \sin \omega_c t + V_n \sin \omega_c t \cos \omega_{\text{diff}}t$$
$$+ V_n \cos \omega_c t \sin \omega_{\text{diff}}t$$
$$= (V_c + V_n \cos \omega_{\text{diff}}t) \sin \omega_c t + V_n \cos \omega_c t \sin \omega_{\text{diff}}t$$
$$= V \sin (\omega_c t + \phi),$$

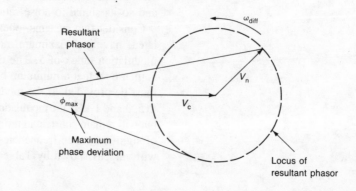

Fig. 2.6 Interference in an f.m. system.

where

$$V = \sqrt{[(V_c + V_n \cos \omega_{\text{diff}}t)^2 + V_n^2 \sin^2 \omega_{\text{diff}}t]}$$
$$= \sqrt{[V_c^2 + V_n^2 + 2V_cV_n \cos \omega_{\text{diff}}t]}$$

and

$$\phi = \tan^{-1}\left[\frac{V_n \sin \omega_{\text{diff}}t}{V_c + V_n \cos \omega_{\text{diff}}t}\right].$$

Usually, $V_c \gg V_n$ and then

$$V \simeq V_c\left[1 + \frac{V_n}{V_c} \cos \omega_{\text{diff}}t\right]$$

and

$$\phi = \tan^{-1}\left[\frac{V_n}{V_c} \sin \omega_{\text{diff}}t\right] \simeq \frac{V_n}{V_c} \sin \omega_{\text{diff}}t.$$

The magnitude V of the resultant voltage varies between the limits $V_c \pm V_n$ so that the f.m. signal is amplitude modulated. The resultant voltage has a sinusoidally varying phase, relative to the unmodulated carrier, with a peak phase deviation of V_n/V_c. The resulting frequency deviation f_{dev} is

$$f_{\text{dev}} = \frac{\omega_{\text{dev}}}{2\pi} = \frac{1}{2\pi} \frac{d}{dt}\left(\frac{V_n}{V_c} \sin \omega_{\text{diff}}t\right)$$

$$= \frac{1}{2\pi}\left(\omega_{\text{diff}} \frac{V_n}{V_c} \cos \omega_{\text{diff}}t\right)$$

or $f_{\text{dev}} = f_{\text{diff}} \frac{V_n}{V_c} \cos \omega_{\text{diff}}t.$ (2.24)

The peak frequency deviation is $f_{\text{diff}}V_n/V_c$ and the r.m.s. frequency deviation is $f_{\text{diff}}V_n/\sqrt{2}V_c$.

When the carrier and noise voltages are at the same frequency their difference frequency is zero and there is no output noise. The f.m. receiver will produce an output that consists solely of the stronger signal, the weaker signal being completely suppressed. This is known as the *capture effect*.

Example 2.5

A 10 mV, 100 MHz carrier has a 25 μV, 100.1 MHz interfering signal superimposed upon it. Calculate the peak phase and frequency deviations of the carrier that are produced.

Solution
The peak phase deviation = $(25 \times 10^{-6})/(10 \times 10^{-3}) = 2.5 \times 10^{-3}$ rad.
(*Ans.*)
The peak frequency deviation = $2.5 \times 10^{-3} \times 0.1 \times 10 = 250$ Hz.
(*Ans.*)

Noise Output Power

The noise power δN_0 at the output of the detector in an f.m. receiver in an audio bandwidth of B Hz is proportional to $V_n^2 f_{diff}^2 / 2 V_c^2$. Now $V_n^2 / 2 = N_0 = P \delta f$, where P is the noise power density spectrum in watts/hertz. Hence

$$\delta N_0 = \frac{KP f_{diff}^2 \delta f}{V_c^2}.$$

(K is the transfer function of the f.m. detector, i.e. an input Δf produces $K \Delta f$ volts at the output.)

The total noise output power in the audio bandwidth is

$$N_0 = \int_{-B}^{B} \frac{KP f_{diff}^2 \mathrm{d}f}{V_c^2} = \frac{2KP}{V_c^2} \int_0^B f_{diff}^2 \mathrm{d}f$$

$$= \frac{2KP}{V_c^2} \left[\frac{f_{diff}^3}{3} \right]_0^B$$

or $\quad N_0 = \frac{2KPB^3}{3V_c^2}.$ \hfill (2.25)

Thus the noise output power is proportional to PB^3/V_c^2.

The Triangular Noise Spectrum

Random noise can be considered to consist of a large number N of equal-amplitude, equally spaced, sinusoidal voltages in a bandwidth of N Hz. The previous analysis can then be applied to demonstrate that an f.m. system has a triangular noise voltage spectrum and a parabolic noise power spectrum.

The output voltage of an f.m. detector is proportional to the frequency deviation of the input voltage. This means (equation (2.24)) that the noise output voltage is proportional to frequency and hence has a triangular spectrum. The output noise voltage rises linearly from zero volts at zero frequency to a maximum, equal to V_n/V_c, at the frequency equal to the rated system deviation f_d (see Fig. 2.7). Usually, the passband of the audio amplifier which follows the detector will be less than the rated system deviation and it will remove all noise voltages at frequencies higher than $f_{m(max)}$. The output noise voltage of an a.m. receiver with the *same* signal and noise voltages would be proportional to V_n/V_c; thus the areas enclosed by the points ADE and ABCD represent, respectively, the output noise voltages of f.m. and a.m. systems having the same audio passband. Area ADE is smaller than area ABCD, which is an indication that the f.m. system has a smaller output noise than the a.m. system. This means that the use of frequency modulation can provide an increase in the output signal-to-noise ratio of a system.

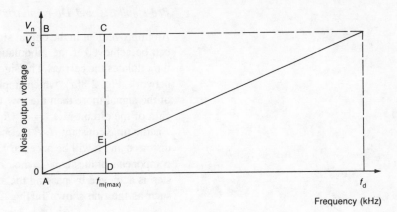

Fig. 2.7 The triangular noise spectrum.

Signal-to-noise Ratio Improvement

The signal output power of an f.m. system is $K(kf_d)^2/2$ and the noise output power is equal to $2KPB^3/3V_c^2$ and so the output signal-to-noise ratio is

$$\left[\frac{\sqrt{3}kf_d}{B}\right]^2 \frac{V_c^2}{4PB} = (\sqrt{3}m_f)^2 \frac{V_c^2}{4PB}.$$

An a.m. system, modulated to a depth of $m\%$, has an output signal-to-noise ratio of

$$\frac{m^2 P_c}{2PB} = \frac{m^2 V_c^2}{4PB}.$$

For a proper comparison of the signal-to-noise ratios assume that both systems are 100% modulated, i.e. $m_f = D$ and $m = 1$. Then

$$\frac{\text{signal-to-noise ratio (f.m.)}}{\text{signal-to-noise ratio (a.m.)}} = 3(D)^2 \tag{2.26}$$

or, in decibels,

$$\text{Signal-to-noise ratio improvement} = 20\log_{10}(\sqrt{3}D). \tag{2.26a}$$

Example 2.6

An a.m. system has an output signal-to-noise ratio of 30 dB. What output signal-to-noise ratio could be obtained if frequency modulation with a deviation ratio of 5 were employed instead and the transmitted power were reduced by 50%?

Solution

From equation (2.26) the improvement in signal-to-noise ratio is $20\log_{10}(\sqrt{3} \times 5) = 18.75$ dB. The 50% reduction in the transmitted power is equivalent to $10\log_{10} 2 = 6$ dB reduction. Therefore, the new signal-to-noise ratio $= 30 + 18.75 - 6 = 42.75$ dB. (*Ans.*)

Pre-emphasis and De-emphasis

An improvement in the output signal-to-noise ratio of an f.m. system can be achieved if the modulating signal is *pre-emphasized* before it modulates the carrier. The signal is passed through a pre-emphasis network, Fig. 2.8(*a*), which amplifies the high-frequency components of the signal more than the low-frequency components. The voltage gain of the circuit is $A_v = g_m(R_L + j\omega L) = g_m R_L(1 + \omega\tau)$, where τ is the time constant, L/R seconds, of the circuit. The signal is then distorted and it will be necessary at the receiver to restore the various components of the signal to their original amplitude relationships. This step is achieved by passing the signal through a de-emphasis circuit such as the one shown in Fig. 2.8(*b*). For this circuit,

$$V_{out} = \frac{V_{in} \cdot 1/j\omega C}{R + 1/j\omega C} = \frac{V_{in}}{1 + j\omega CR}$$

and so

$$\left|\frac{V_{out}}{V_{in}}\right| = \frac{1}{\sqrt{(1 + \omega^2 R^2 C^2)}}.$$

This circuit will have a 3 dB loss when

$$\left|\frac{V_{out}}{V_{in}}\right| = \frac{1}{\sqrt{2}} = \frac{1}{\sqrt{(1 + \omega_1^2 R^2 C^2)}},$$

i.e. at the frequency f_1, where

$$f_1 = \frac{1}{2\pi RC} \text{ Hz}.$$

Thus

$$\left|\frac{V_{out}}{V_{in}}\right|^2 = \frac{1}{1 + f^2/f_1^2}.$$

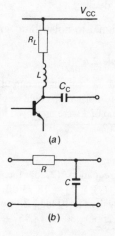

(a)

(b)

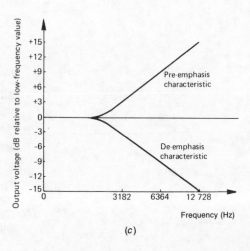

(c)

Fig. 2.8 (*a*) A pre-emphasis circuit, (*b*) a de-emphasis circuit, and (*c*) pre-emphasis and de-emphasis characteristics for 50 μs time constant.

At frequency $f = 2f_1$

$$\left|\frac{V_{\text{out}}}{V_{\text{in}}}\right|^2 = \frac{1}{1 + 4} \text{ or } -7 \text{ dB};$$

at frequency $f = 4f_1$

$$\left|\frac{V_{\text{out}}}{V_{\text{in}}}\right|^2 = \frac{1}{1 + 16} = -12 \text{ dB},$$

and so on. Therefore at higher frequencies than f_1 the attenuation of the network falls at 6 dB/octave. To ensure that the component frequencies are restored to their original amplitude relationships the time constants of the pre-emphasis and de-emphasis networks must be equal. In the UK sound broadcast system a time constant $\tau = L/R = CR = 50 \ \mu\text{s}$ is used. Then

$$f_1 = \frac{1}{2\pi \times 50 \times 10^{-6}} = 3183 \text{ Hz},$$

giving the pre-emphasis and de-emphasis characteristics shown in Fig. 2.8(c).

To determine the effectiveness of pre-/de-emphasis in increasing the output signal-to-noise ratio of an f.m. system it must be compared with an a.m. system with the same value of $f_{\text{m(max)}}$. Assume a 1 Ω resistance; then for the a.m. system the output noise voltage V_{no} is constant with frequency and hence

$$\text{a.m. noise power} = \int_0^{f_{\text{m(max)}}} V_{\text{no}}^2 \frac{\text{d}f}{1 + (f/f_1)^2}.$$

Using a standard integral

$$\int \frac{\text{d}x}{1 + a^2 x^2} = \frac{1}{a} \tan^{-1}(ax),$$

this gives $V_{\text{no}}^2 f_1 [\tan^{-1}(f/f_1)]_0^{f_{\text{m(max)}}}$. Therefore,

$$N_{\text{o(a.m.)}} = V_{\text{no}}^2 f_1 \tan^{-1}\left[\frac{f_{\text{m(max)}}}{f_1}\right]. \tag{2.27}$$

The noise output voltage of an f.m. system is proportional to frequency, i.e. $V_{\text{no}} f/f_{\text{d}}$. Hence

$$\text{f.m. noise power} = V_{\text{no}}^2 \int_0^{f_{\text{m(max)}}} \left[\left(\frac{f}{f_{\text{d}}}\right)^2 \frac{\text{d}f}{1 + (f/f_1)^2}\right]$$

$$= \frac{V_{\text{no}}^2}{f_{\text{d}}^2} \int_0^{f_{\text{m(max)}}} \frac{f^2 \text{d}f}{1 + (f/f_1)^2}.$$

Using another standard integral

$$\int \frac{x^2 \text{d}x}{1 + a^2 x^2} = x^3[ax - \tan^{-1}(ax)],$$

gives

$$\text{f.m. noise power} = \frac{V_{no}^2 f_1^3}{f_d^2} \left[\frac{f_{m(max)}}{f_1} - \tan^{-1} \left(\frac{f_{m(max)}}{f_1} \right) \right].$$

(2.28)

If $f_1 = 3183$ Hz, $f_{m(max)} = 15$ kHz, and $f_d = 75$ kHz, the ratio

$$\frac{\text{a.m. noise power}}{\text{f.m. noise power}} = \frac{4334}{19.2} = 225.68 \text{ or } 23.54 \text{ dB}.$$

Without pre-emphasis the signal-to-noise ratio improvement of f.m. over a.m. was 18.75 dB so that the increase provided by the use of pre-emphasis is 23.54 − 18.75 or 4.8 dB.

Frequency Modulators

A frequency modulator converts an input signal voltage into an output frequency change. The change in the carrier frequency will be equal to the rated system deviation when the signal voltage is at its maximum value. The *sensitivity* of the modulator is equal to the ratio $f_d/V_{m(max)}$ in kHz/V. Ideally, there should be no amplitude modulation of the carrier. Frequency modulators are classified as being either *direct* or *indirect* types.

Direct-frequency modulators

Figure 2.9 illustrates the basic principle upon which the majority of direct-frequency modulators are based. The frequency of oscillation of an oscillator is determined by a parallel-tuned circuit whose total capacitance (or, much less often, its total inductance) is provided by the parallel connection of a physical capacitor C_1 and a voltage-variable reactance. The modulating signal voltage is applied, usually together with a bias voltage, to the voltage-variable reactance to vary its capacitance. The change in capacitance will then alter the resonant frequency of the tuned circuit and so vary the frequency of oscillation. In this way the modulating signal voltage is able to modulate the carrier frequency.

When the modulating signal voltage is zero the voltage-variable reactance has its average value C_e of capacitance and the oscillation

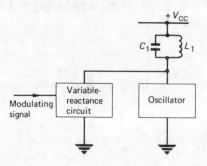

Fig. 2.9 Principle of a frequency modulator.

frequency will be the unmodulated carrier frequency, i.e.

$$f_c \simeq \frac{1}{2\pi\sqrt{[L_1(C_1 + C_e)]}} \text{ Hz} = \frac{1}{2\pi\sqrt{(L_1 C_T)}}. \tag{2.29}$$

When the modulating signal voltage is instantaneously equal to v_m volts the total capacitance of the tuned circuit is varied by an amount δC_T. Then

$$f_c + \delta f_c = \frac{1}{2\pi\sqrt{[L_1(C_T + \delta C_T)]}}. \tag{2.30}$$

Divide equation (2.30) by (2.29) to give

$$1 + \frac{\delta f_c}{f_c} = \sqrt{\left(\frac{C_T}{C_T + \delta C_T}\right)} = \frac{1}{\sqrt{(1 + \delta C_T/C_T)}}. \tag{2.31}$$

Since $\delta C_T \ll C_T$ equation (2.31) can be written as

$$1 + \frac{\delta f_c}{f_c} \simeq 1 - \frac{\delta C_T}{2 C_T}$$

or

$$\frac{\delta f_c}{f_c} \simeq \frac{-\delta C_T}{2 C_T}. \tag{2.32}$$

This means that a fractional increase in C_T will produce a fractional decrease in f_c which is approximately half as large.

Reactance Frequency Modulator

The circuit of a reactance frequency modulator that employs a bipolar transistor is shown in Fig. 2.10. R_1, R_2, R_3 and C_1 are bias and decoupling components, L_1 is an r.f. choke and C_3 is a d.c. block. The required reactance effect is provided by C_2 and R_4. The current i flowing in the series $C_2 R_4$ circuit is $i = V_{ce}/(R_4 + 1/j\omega C_2)$ and this develops a voltage

$$V_{in} = \frac{V_{ce} j\omega C_2 R_4}{1 + j\omega C_2 R_4}$$

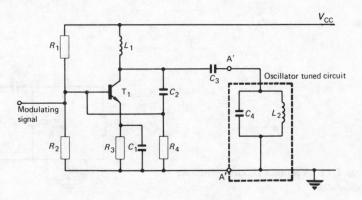

Fig. 2.10 Reactance frequency modulator.

across R_4. The collector current i_c of T_1 is equal to $g_m V_{in}$ and so the output admittance Y_{out} of the modulator is

$$Y_{out} = \frac{i_c}{V_{ce}} = \frac{j\omega g_m C_2 R_4}{1 + j\omega C_2 R_4} \simeq j g_m \omega C_2 R_4. \qquad (2.33)$$

The circuit thus provides an effective capacitance of $C_e = g_m C_2 R_4$ across the oscillator's tuned circuit that is directly proportional to the mutual conductance g_m of the transistor. Since g_m can be varied by applying the modulating signal to the input terminals of the circuit, frequency modulation of the carrier is achieved.

Example 2.7

The reactance frequency modulator of Fig. 2.10 has $C_2 = 5$ pF and $R_4 = 1$ kΩ. The total tuned circuit capacitance when there is no input signal is 500 pF and the unmodulated carrier frequency is 1 MHz. Calculate the deviation of the carrier frequency when a modulating signal varies the mutual conductance of T_1 by 2 mS.

Solution

$$\delta C = 2 \times 10^{-3} \times 5 \times 10^{-12} \times 10^3 = 10 \text{ pF}.$$

From equation (2.32),

$$\delta f_c = \frac{-1 \times 10^6 \times 10}{2 \times 500} = -10 \text{ kHz}. \qquad (Ans.)$$

Varactor Diode Frequency Modulator

The basic circuit of a varactor diode frequency modulator is shown in Fig. 2.11. The varactor (or voltage-variable) diode is connected in parallel with the tuned circuit $C_1 L_1$ of the oscillator to be modulated. C_2 is a d.c. blocking component while L_2 is an r.f. choke. When the modulating signal voltage is zero the varactor diode is reverse-biased by the bias voltage V_B volts and it then has a capacitance of $C_d = C_0/\sqrt{V_B}$ pF, where C_0 is the capacitance for zero applied voltage. The frequency of oscillation is then

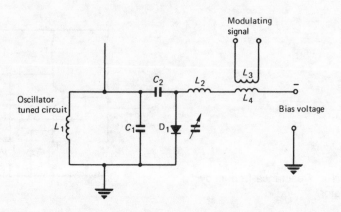

Fig. 2.11 Varactor diode frequency modulator.

$$f_c = \frac{1}{2\pi\sqrt{[L(C_1 + C_d)]}} \text{ Hz.}$$

When a modulating signal is applied to the circuit the reverse-bias voltage becomes $V_B + V_m \sin \omega_m t$ and

$$C_d + \delta C_d = \frac{C_0}{\sqrt{(V_B + V_m \sin \omega_m t)}}$$

$$= \frac{C_0}{\sqrt{V_B}\sqrt{\left(1 + \frac{V_m}{V_B} \sin \omega_m t\right)}}.$$

$$C_d + \delta C_d = \frac{C_d}{\sqrt{\left(1 + \frac{V_m}{V_B} \sin \omega_m t\right)}}.$$

Therefore,

$$1 + \frac{\delta C_d}{C_d} = \frac{1}{\sqrt{\left(1 + \frac{V_m}{V_B} \sin \omega_m t\right)}} \simeq 1 - \frac{V_m}{2V_B} \sin \omega_m t$$

and

$$\delta C_d = \frac{-V_m}{2V_B} C_d \sin \omega_m t. \tag{2.34}$$

Equation (2.34) can be substituted into equation (2.32) to obtain an expression for the deviation of the carrier frequency. Better linearity, with a consequent reduction in signal distortion, is obtained if two back-to-back varactor diodes are employed (see Fig. 2.12). The effective capacitances of the two diodes are now in series and so each must have *twice* the capacitance of the single varactor diode they replace.

Example 2.8

The circuit of Fig. 2.12 is a part of a frequency modulator which operates at a centre frequency of 1 MHz. $L_1 = 100 \ \mu H$, $C_1 = 200$ pF and each varactor diode has $C_0 = 200$ pF. Calculate the values of the bias voltage and the peak modulating voltage if the frequency deviation is 1 kHz. Also calculate the sensitivity of the modulator.

Solution
The necessary tuning capacitance is equal to

$$\frac{1}{4\pi^2 \times 10^{12} \times 100 \times 10^{-6}} \simeq 253 \text{ pF}$$

and so the varactor diodes must contribute 53 pF. Each diode must therefore

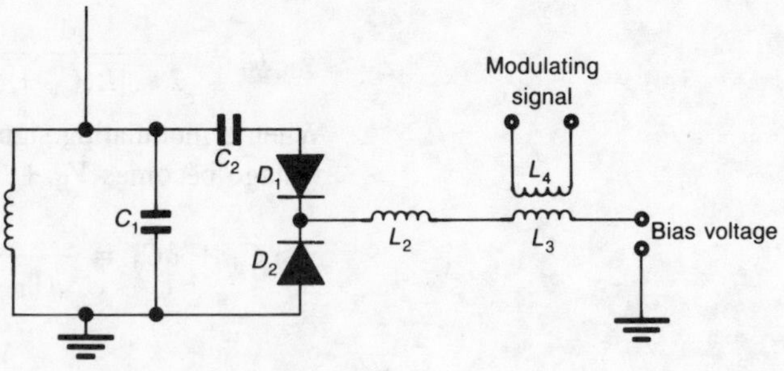

Fig. 2.12 Improved varactor diode frequency modulator.

have a capacitance of 106 pF. This means that the reverse bias voltage V_B must be

$$V_B = (200/106)^2 = 3.56 \text{ V}. \quad (Ans.)$$

When the carrier frequency is deviated by 1 kHz

$$\frac{1 \times 10^3}{1 \times 10^6} = \frac{\delta C_T}{2 \times 253 \times 10^{-12}} \text{ or } \delta C_T = 0.5 \text{ pF}$$

and then C_T is either 253.5 pF or 252.5 pF. Each varactor diode must then have a capacitance of either 107 pF or 105 pF. Choosing the former $V_B - V_m = (200/107)^2 = 3.49$ V and the peak modulating voltage is $3.56 - 3.49 = 70$ mV. $(Ans.)$

On the other half cycle, $V_B + V_m = (200/105)^2 = 3.63$ V and the peak signal voltage is $3.63 - 3.56 = 70$ mV. $(Ans.)$

The sensitivity of the modulator is $1/(70 \times 10^{-3}) = 14.286$ kHz/V.

$(Ans.)$

The varactor diode and reactance frequency modulators produce n.b.f.m. signals because $\delta C_T \ll C_T$ and if wideband modulation is wanted some stages of frequency multiplication will be necessary.

Indirect Frequency Modulation

The inherent stability of the unmodulated carrier frequency when a direct modulator is used is not good enough to meet modern standards. There are two ways in which this difficulty can be overcome: (a) direct frequency modulation with automatic frequency control applied to the transmitter; and (b) indirect frequency modulation of a very stable crystal oscillator.

The expressions for the instantaneous frequency and voltage of a sinusoidally modulated phase-modulated wave are given by equations (2.22) and (2.23), respectively. If the modulating signal $v_m = V_m \sin \omega_m t$ is integrated before it is applied to the phase modulator, to give $V_m/\omega_m \cos \omega_m t$, these equations become $f_i = f_c + k\Phi_d \sin \omega_m t$ and

$$v_c = V_c \sin \left(\omega_c t + \frac{k\Phi_d}{\omega_m} \cos \omega_m t \right).$$

The frequency deviation is now proportional to the voltage of the modulating signal only, and inversely proportional to the modulating signal frequency. These are, of course, the characteristics of a frequency-modulated waveform. Hence, an f.m. wave can be obtained by integrating the modulating signal and then using a phase modulator.

Phase Modulators

A phase-modulated wave can be generated using an a.m. balanced modulator and a carrier source that has both in-phase and quadrature outputs. Figure 2.13 shows the circuit of an Armstrong phase modulator. The balanced modulator has two inputs: the modulating signal and a 90° phase-shifted carrier voltage and it generates the upper and lower sidebands of amplitude modulation but suppresses the carrier. Assuming the modulating signal to be of sinusoidal waveform, the output of the balanced modulator is $mV_c \cos \omega_c t \sin \omega_m t$. This is added to the non-phase shifted carrier voltage to give an output of

$$V_{out} = V_c \sin \omega_c t + mV_c \cos \omega_c t \sin \omega_m t$$
$$= \sqrt{(V_c^2 + m^2 V_c^2 \sin^2 \omega_m t)} \sin (\omega_c t + \theta),$$

where $\theta = \tan^{-1} (m \sin \omega_m t)$. Hence,

$$V_{out} \simeq V_c \sin (\omega_c t + m \sin \omega_m t)$$

which is of the same form as equation (2.22).

The carrier can be produced by a stable crystal oscillator. The maximum modulation index that this circuit can give is only about 0.25 and so several stages of frequency multiplication will be necessary to obtain a wideband f.m. wave. If, for example, the modulation index is 0.25 and it is to be equal to 5, frequency multiplication of 20 times is required. The crystal oscillator frequency should then, of course, be equal to f_c/N, where f_c is the wanted carrier frequency and N is the multiplication ratio.

Example 2.9

An f.m. transmitter has an unmodulated carrier frequency of 90 MHz and a rated system deviation of 75 kHz. The Armstrong modulator used can only produce a peak frequency deviation of 3 kHz. Calculate (a) the multiplication ratio used, and (b) the frequency of the crystal oscillator.

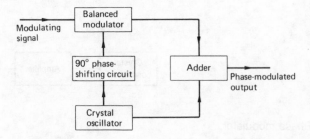

Fig. 2.13 Armstrong phase modulator.

Solution

(*a*) $N = 75/3 = 25$.   (*Ans.*)

(*b*) Oscillator frequency $= 90/25 = 3.6$ MHz.   (*Ans.*)

An alternative form of phase modulator is shown in Fig. 2.14. The buffered output of a crystal oscillator is applied to a parallel-tuned circuit whose resonant frequency can be varied by applying the modulating signal to the varactor diode. The output voltage of the modulator will then be phase modulated. Capacitor C_1 is an r.f. bypass and C_2 is a d.c. block.

Example 2.10

The varactor diode used in the circuit of Fig. 2.14 has the capacitance-voltage characteristic $C_d = 200/\sqrt{V}$ pF. When the d.c. reverse-bias voltage is 4 V the circuit is resonant at 5 MHz and has a Q-factor of 20. Calculate the peak phase deviation produced when a 50 mV peak sinusoidal modulating signal is applied to the circuit.

Solution

When the modulating signal voltage V_m is zero $C_d = 200/\sqrt{4} = 100$ pF. When the modulating signal is applied to the circuit the minimum and maximum values of C_d are

$$C_{d(min)} = 200/\sqrt{4.05} = 99.38 \text{ pF}$$

and $C_{d(max)} = 200/\sqrt{3.95} = 100.63$ pF. Hence δC is either 0.62 pF or it is 0.63 pF; using the mean value in $\delta f_c/f_c = -\delta C/2C$ gives

$$\delta f_c = \frac{5 \times 10^6 \times 0.625}{200} = 15.625 \text{ kHz.}$$

The impedance Z of the tuned circuit at any frequency is given by

$$Z = \frac{R_d}{1 + jQB/f_0},$$

where R_d is the dynamic impedance, Q is the Q-factor, B is the bandwidth considered and f_0 is the resonant frequency. The angle θ of the impedance is $\pm\tan^{-1}(QB/f_0)$ or

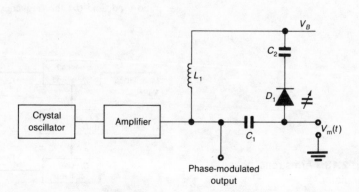

Crystal oscillator — Amplifier — L_1 — C_1 — C_2 — D_1 — V_B — $V_m(t)$

Phase-modulated output

Fig. 2.14 Phase modulator.

$$\theta = \pm \tan^{-1} \left[\frac{20 \times 31\ 250}{5 \times 10} \right] = \pm 0.124 \text{ rad.} \quad (Ans.)$$

Frequency-modulation Detectors

A frequency-modulation detector is required to produce an output voltage that is directly proportional to the instantaneous frequency of its input signal. It has a transfer function with units of V/kHz. Before the widespread use of integrated circuits in radio receivers, most f.m. detectors were based upon the principle of first converting the f.m. signal into a signal whose amplitude varied in proportion to the frequency deviation and then envelope-detecting in the converted signal. The concept is illustrated, in block diagram form, by Fig. 2.15. The two most commonly employed examples of this technique are the phase discriminator and the ratio detector (see *Radio Systems for Technicians*).

Most modern radio receivers make full use of the widely available ICs and these receivers are most likely to employ either a *quadrature detector* or a *phase-locked loop* (p.l.l.) (see *Electronics IV*).

The Quadrature Detector

The f.m. signal to be detected is split into two parts. One part is directly applied to one of the two inputs of an analogue multiplier circuit. The other part is passed through a capacitor having a high reactance at the unmodulated carrier frequency and then to a parallel-tuned circuit, see Fig. 2.16. The capacitor C_1 introduces a phase shift of very nearly 90° and the tuned circuit a phase shift that depends upon frequency. The impedance of the tuned circuit is

$$Z = \frac{R_d}{1 + jQB/f_0}$$

At the resonant frequency $B = 0$ and the circuit is purely resistive; at any frequency off of resonance the circuit introduces a phase shift of $\pm \tan^{-1} [QB/f_0]$. The voltage across the tuned circuit is applied to the other input of the multiplier and an output proportional to the product of the two input signals is obtained. This output is passed through a low-pass filter to obtain the original modulating signal $V_m(t)$.

If the unmodulated carrier frequency is $\omega_c/2\pi$ and the instantaneous frequency of the input signal is $\omega = \omega_c \pm \delta\omega_c$ then the phase shift θ between the two multiplier inputs is

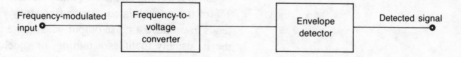

Fig. 2.15 Principle of an f.m. detector.

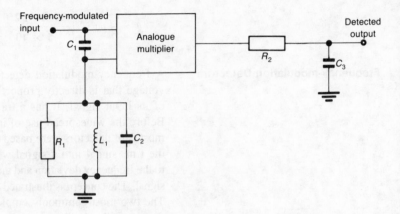

Fig. 2.16 Quadrature detector.

$$\theta = (\pi/2) \pm \tan^{-1} [2Q\delta\omega_c/\omega_c] = (\pi/2) \pm \phi.$$

(Plus sign if $\omega > \omega_c$, negative sign if $\omega < \omega_c$.) If the input signal is $V \sin \omega t$ the phase-shifted signal will be $V \sin (\omega t + (\pi/2) \pm \phi)$, or $V \cos (\omega t \pm \phi)$. Multiplied together these two signals give

$$V^2 \sin \omega t \cos (\omega t \pm \phi) = \frac{V^2}{2} [\sin (2\omega t \pm \phi)$$
$$+ \sin (\pm \phi)]$$

(using the identity $2 \sin A \cos B = \sin (A + B) + \sin (A - B)$).

The output of the low-pass filter is

$$\frac{V^2}{2} \sin (\pm \phi) = \frac{V^2}{2} \sin \left(\pm \tan^{-1} \frac{2Q\delta\omega_c}{\omega_c} \right).$$

For $2Q\delta\omega_c \ll \omega_c$,

$$\sin \left(\tan^{-1} \frac{2Q\delta\omega_c}{\omega_c} \right) \simeq \frac{2Q\delta\omega_c}{\omega_c}.$$

This output voltage bears a linear relationship to $\delta\omega_c$, i.e. to the frequency deviation of the carrier, and it is therefore the desired modulating signal.

The quadrature detector is widely employed in modern radio receivers and it is usually incorporated within an IC that also performs several other circuit functions. The detector can operate with small amplitude signals (100 μV or so) and it is easy to set up; it is only necessary to tune the phase-shifting circuit to the incoming carrier frequency. The circuit provides good linearity, with consequent small distortion, as long as the frequency deviation is not more than about 1% of the unmodulated carrier frequency. The quadrature detector tends to produce a noise output when there is no input signal and so the IC usually includes a muting, or squelch, circuit, too.

If the analogue multiplier is replaced by an AND gate and the input signal is hard limited to produce a variable-frequency pulse waveform, a *coincidence detector* is obtained.

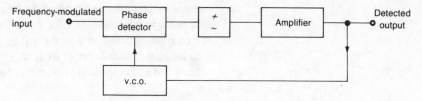

Frequency-modulated input

Phase detector

÷

Amplifier

Detected output

v.c.o.

Fig. 2.17 Phase-locked-loop detector.

The Phase-locked-loop Detector

The phase-locked loop (p.l.l.) can be operated as a detector of f.m. signals when it has more or less the same performance as the quadrature detector. The p.l.l. detector does not need an externally connected tuned circuit but, on the other hand, an external piezo-electric crystal is required. In some cases the 10.7 MHz output of an i.f. amplifier may be converted to a lower frequency in the region of 200 kHz before it is applied to the p.l.l. detector. Sometimes the mixer used is on-chip with the p.l.l.

The block schematic diagram of a p.l.l. f.m. detector is shown in Fig. 2.17. The free-running frequency f_0 of the voltage-controlled oscillator (v.c.o.) is set to be equal to the unmodulated carrier frequency f_c of the signal to be demodulated. The phase detector generates an output voltage that is directly proportional to the phase difference, or *phase error*, between the f.m. signal and the v.c.o. voltage. This error voltage is passed through a low-pass filter and then amplified to produce both the output voltage and a control voltage for the v.c.o. The free-running frequency of the v.c.o. is varied by the voltage applied to its control terminal. The polarity of the control voltage is always such that it varies the frequency of the v.c.o. in the direction that reduces the difference frequency $\Delta f = f_c - f_0$ and hence the phase error. This means that the frequency of the v.c.o. is forced to try to become equal to the instantaneous frequency of the input signal.

Suppose that, initially, the input signal is not modulated. The action of the p.l.l. reduces the difference between the input carrier frequency f_c and the frequency f_0 of the v.c.o. Once the v.c.o. frequency has moved to be very close to the carrier frequency the loop attains *lock*. Then the v.c.o. rapidly attains the same frequency as the input signal but there is *always* a phase difference between the two voltages. This phase error must always exist in order to maintain the v.c.o. control voltage at the required value. This voltage, and hence the output voltage, is a d.c. voltage.

When the input signal is frequency modulated a similar action takes place. As the input signal frequency deviates from the unmodulated carrier frequency the error voltage varies too to ensure that the v.c.o. tracks to minimize the phase error. As a result the instantaneous v.c.o. frequency is always approximately equal to the frequency of the incoming signal. The output of the low-pass filter is the detected modulating signal voltage. The filter must have a cut-off frequency equal to the maximum modulating frequency $f_{m(max)}$ of the f.m. signal in order to minimize the effects of noise and interference. Should the free-running frequency of the v.c.o. *not* be set to be exactly equal to the unmodulated carrier frequency the detected output voltage will have a d.c. component, but this is (usually) unimportant.

Analysis

Once the loop is in *lock* the frequency of the v.c.o. is equal to the carrier frequency. The f.m. input signal is

$$v_S = V_S \sin [\omega_c t + 2\pi k f_d \int_0^t V_m(t)]$$
$$= V_S \sin [\omega_c t + K_S \int_0^t V_m(t)\, dt] = V_S \sin [\omega_c t + \theta_1(t)].$$

The output voltage of the v.c.o. is

$$v_0 = V_0 \cos [\omega_c t + \theta_2(t)].$$

These two signals are the inputs to the phase detector which generates an output voltage v_d that is directly proportional to their product. Thus

$$v_d = K_D V_S V_0 \sin [\omega_c t + \theta_1(t)] \cos [\omega_c t + \theta_2(t)]$$
$$= \frac{K_D V_S V_0}{2} \{ \sin [\theta_1(t) - \theta_2(t)]$$
$$+ \sin [2\omega_c t + \theta_1(t) + \theta_2(t)] \}.$$

K_D is the gain factor, in V/rad, of the phase detector. The low-pass filter will only transmit the lower-frequency components of v_d so that the voltage appearing at the output of the circuit is

$$v_0 = \frac{K_D A_V V_S V_0}{2} \sin [\theta_1(t) - \theta_2(t)].$$

Since the loop is locked the phase error will be small and

$$\sin [\theta_1(t) - \theta_2(t)] \simeq \theta_1(t) - \theta_2(t),$$

giving

$$v_0 = \frac{K_D A_V V_S V_0}{2} [\theta_1(t) - \theta_2(t)]. \qquad (2.35)$$

The instantaneous angular velocity ω_0^1 of the v.c.o. output voltage is $\omega_0^1 = \omega_0 + K_0 v_0(t)$, where ω_0 is the free-running angular velocity and K_0 is the conversion gain of the v.c.o. The conversion gain K_0 relates the frequency of the v.c.o. to its control voltage and it is expressed in rad/s/volt or in kHz/volt. Since $\omega = d\theta/dt$, $d\theta_2/dt = K_0 v_0(t)$; also

$$d\theta_1/dt = K_S V_m(t).$$

Equation (2.35) can be written as

$$V_0(t) = \frac{K_D A_V V_S V_0}{2} \left[\theta_1(t) - K_0 \int_0^t v_0(t)\,\mathrm{d}t\right],$$

and differentiating with respect to time,

$$\frac{\mathrm{d}v_0(t)}{\mathrm{d}t} = \frac{K_D A_V V_S V_0}{2} \left[\frac{\mathrm{d}\theta_1(t)}{\mathrm{d}t} - K_0 v_0(t)\right]$$

or $\dfrac{\mathrm{d}\theta_1(t)}{\mathrm{d}t} = \dfrac{2\mathrm{d}v_0(t)}{\mathrm{d}t(K_D A_V V_S V_0)} + K_0 v_0(t) \simeq K_0 v_0(t).$

Therefore,

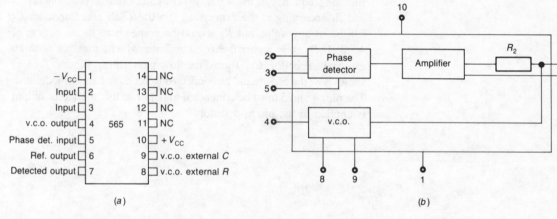

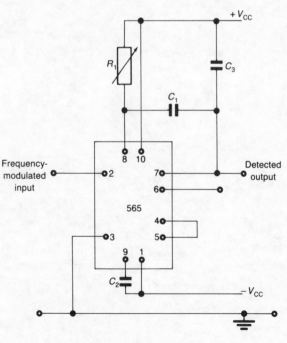

Fig. 2.18 The 565 p.l.l. IC: (a) pin connections, (b) internal block diagram, (c) connection as an f.m. detector.

$$K_S V_m(t) = K_0 v_0(t),$$

or $\qquad v_0(t) = \dfrac{K_S}{K_0} V_m(t).$ \hfill (2.36)

The ratio K_S/K_0 has the dimensions of kHz/kHz/V or volts and so the output of the detector is the wanted modulating signal $V_m(t)$.

The 565 as a Frequency Demodulator

The 565 IC is a widely used p.l.l. for operation at frequencies up to about 500 kHz. Figure 2.18(a) and (b) show, respectively, the pin connections and the internal block diagram of the 565. The free-running frequency of the v.c.o. is set by the external components C_1 and R_1 according to the expression $f = 0.3/C_1 R_1$ Hz. Capacitor C_1 can be of any value but R_1 should be somewhere in the region of 4000 Ω. R_2 is an internal 3.6 kΩ resistor which, together with an external capacitor C_2, forms the low-pass filter. Figure 2.18(c) shows how the 565 should be connected to act as an f.m. demodulator. The pins 4 and 5 must be connected together so that the v.c.o. output is applied to the phase detector.

3 Transmission Lines

A transmission line consists of two conductors separated from one another by a dielectric; the two conductors may be identical and form a twin line or one conductor may be positioned inside the other (a coaxial line). A transmission line provides a means for guiding an electromagnetic wave from one point to another. The behaviour of a transmission line is normally described in terms of the current and voltage waves which propagate along it. The performance is determined by the values of its *secondary coefficients*. These are: the *characteristic impedance* Z_0, the *propagation coefficient* γ, and the velocity of propagation v. These, in turn, are functions of the *primary coefficients* of the line, i.e. the resistance R, the conductance G, the inductance L and the capacitance C. The full expressions for the secondary coefficients involve all four of the primary coefficients and also the frequency (*Electrical Principles IV*), but at radio frequencies these expressions can be reduced to simpler, and more convenient, forms. At radio frequencies the characteristic impedance Z_0 of a line is purely resistive and is given by $Z_0 = \sqrt{(L/C)}$ Ω. The propagation coefficient γ is given by

$$\gamma = \frac{R}{2Z_0} + \frac{GZ_0}{2} + j\omega\sqrt{LC}.$$

The real part of γ is known as the *attenuation coefficient* α in nepers/metre (or dB/m), and the imaginary part is the *phase-change coefficient* β in rad/m. The attenuation coefficient is not a constant quantity but instead it increases with increase in frequency; this is because the resistance R is proportional to $\sqrt{\text{frequency}}$ and the conductance G is directly proportional to frequency. Usually,

$$\frac{R}{2Z_0} \gg \frac{GZ_0}{2}$$

and then the dielectric loss can be neglected. The attenuation coefficient α is then equal to the conductor loss $R/2Z_0$ nepers per metre.

The attenuation of an r.f. line is proportional to $\sqrt{\text{frequency}}$ while the wavelength is inversely proportional to frequency. This means that the attenuation per wavelength *decreases* with increase in frequency. At the higher radio frequencies, in particular, the electrical length of a line is often small, perhaps only a fraction of a wavelength, and then the line loss will be small. Such lines are often described as being *low-loss* or even, if the losses are small enough to be neglected, as *loss-free*.

At radio frequencies the phase and group velocities are of equal value and given by

$$v = \frac{\omega}{\beta} = \frac{1}{\sqrt{LC}} \text{ m/s.}$$

At the lower radio frequencies twin conductors may be employed since they are generally less expensive, but at higher frequencies their losses may be excessively high because of radiated energy. At the higher radio frequencies, therefore, coaxial lines are employed. If an air-spaced line has a characteristic impedance of Z_0 then the use of a continuous dielectric of relative permittivity ϵ_r will reduce the impedance to $Z_0/\sqrt{\epsilon_r}$. When the inner conductor of a coaxial line is held in position by spacing discs of thickness t and relative permittivity ϵ_r, spaced distance d apart, the characteristic impedance is equal to

$$\frac{Z_0}{\sqrt{\left[1 + \frac{(\epsilon_r - 1)t}{d}\right]}}.$$

Example 3.1

The inner of a coaxial cable is supported by 0.64 cm thick discs spaced 5 cm apart. The disc material has a relative permittivity of 2.3 and the air-spaced value of the characteristic impedance is 70 ohms. Calculate the characteristic impedance of the cable.

Solution

$$Z_0 = \frac{70}{\sqrt{\left[1 + \frac{(2.3 - 1)\, 0.64}{5}\right]}} = 64.8 \ \Omega. \qquad (Ans.)$$

When a transmission line is used to transmit energy from one point to another the load impedance is chosen, as far as possible, to match the characteristic impedance of the line. This will ensure both the maximum transfer of energy to the load and the absence of standing waves on the line. The behaviour of a matched transmission line is fairly straightforward and an understanding of it is assumed in this chapter (see *Electrical Principles IV*).

Mismatched Transmission Lines

Whenever the load terminals of an r.f. line are closed in an impedance that is not equal to the characteristic impedance of the line, the load will be unable to absorb all of the incident power. A fraction of the incident power will be *reflected* by the load and transmitted back towards the sending end of the line. If the sending-end terminals are

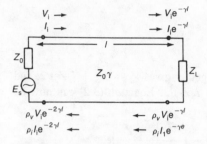

Fig. 3.1 Currents and voltages on a mismatched transmission line.

matched to the source impedance all of the reflected power will be dissipated in the source impedance. If, however, the sending-end terminals are also mismatched some of the returning energy will be further reflected and *multiple* reflections will take place.

Figure 3.1 shows a line of length l metres, having secondary coefficients Z_0 and γ, which is terminated in a load impedance Z_L that is *not* equal to Z_0. The sending-end terminals of the line are matched. When the source is first connected to the line the impedance presented to it is the characteristic impedance of the line Z_0. The *incident* current I_i and voltage V_i into the line are therefore equal to $E_S/2Z_0$ and $E_S/2$, respectively. The incident current and voltage waves propagate along the line and experience both attenuation and phase change as they travel. At the distant end of the line the magnitudes of the incident current and voltage waves are $I_i\mathrm{e}^{-\gamma l}$ and $V_i\mathrm{e}^{-\gamma l}$, respectively.

Since the load impedance Z_L is *not* equal to the characteristic impedance Z_0 both the incident current and the incident voltage waves are reflected. The values of the reflected waves are determined by the current, and voltage, reflection coefficients ρ_i and ρ_v, respectively. The reflected current is equal to $\rho_i I_i\mathrm{e}^{-\gamma l}$ and the reflected voltage is $\rho_v V_i\mathrm{e}^{-\gamma l}$. The reflected power is, of course, equal to the product of the reflected current and the reflected voltage.

The reflected waves propagate along the line towards its sending end and are further attenuated and phase shifted as they travel. At the sending end of the line the reflected current is $\rho_i I_i\mathrm{e}^{-2\gamma l}$ and the reflected voltage is $\rho_v V_i\mathrm{e}^{-2\gamma l}$. Since the source impedance is equal to the characteristic impedance of the line no further reflections occur. At any point along the line the total current and voltage are the phasor sum of the incident and the reflected waves. At a distance x from the sending end of the line

$$V_x = V_i\mathrm{e}^{-\gamma x} + \rho_v V_i\mathrm{e}^{-\gamma(2l-x)}. \tag{3.1}$$

This can be written as

$$V_x = V_i\mathrm{e}^{-\gamma x} + V_r\mathrm{e}^{\gamma x} \tag{3.2}$$

where V_r is the reflected voltage at the receiving end of the line, i.e. $V_r = \rho_v V_i\mathrm{e}^{-\gamma l}$. Similarly, the current at any point distance x from the sending-end terminals is

$$I_x = I_i\mathrm{e}^{-\gamma x} + I_r\mathrm{e}^{\gamma x} \tag{3.3}$$

or $\quad I_x = \dfrac{V_i\mathrm{e}^{-\gamma x}}{Z_0} - \dfrac{V_r\mathrm{e}^{\gamma x}}{Z_0}. \hfill (3.4)$

The minus sign is necessary because the reflected current is *always* in antiphase with the reflected voltage.

It is sometimes convenient to be able to express the current and voltage on a mismatched line in terms of the hyperbolic functions $\cosh x = (\mathrm{e}^x + \mathrm{e}^{-x})/2$ and $\sinh x = (\mathrm{e}^x - \mathrm{e}^{-x})/2$. At the sending end of the line $x = 0$ and, from equations (3.2) and (3.4),

$$V_S = V_i + V_r$$

$$\text{and } I_S = \frac{V_i}{Z_0} - \frac{V_r}{Z_0}.$$

Adding these two equations together gives $V_i = (V_S + I_S Z_0)/2$, and subtracting gives $V_r = (V_S - I_S Z_0)/2$. Equation (3.2) can now be written as

$$V_x = \left[\frac{V_S + I_S Z_0}{2}\right] e^{-\gamma x} + \left[\frac{V_S - I_S Z_0}{2}\right] e^{\gamma x}$$

$$= V_S \left[\frac{e^{\gamma x} + e^{-\gamma x}}{2}\right] - I_S Z_0 \left[\frac{e^{\gamma x} - e^{-\gamma x}}{2}\right]$$

or $V_x = V_S \cosh \gamma x - I_S Z_0 \sinh \gamma x.$ (3.5)

Similarly

$$I_x = I_S \cosh \gamma x - \frac{V_S}{Z_0} \sinh \gamma x. \tag{3.6}$$

Equations (3.5) and (3.6) are known as the *general line equations*.
  The current and voltage at any point on a mismatched line may also be expressed in terms of the current and the voltage at the load terminals. At a distance x from the load terminals

$$V_x = V_I e^{\gamma x} + V_R e^{-\gamma x} \tag{3.7}$$

$$I_x = \frac{V_I e^{\gamma x}}{Z_0} - \frac{V_R e^{-\gamma x}}{Z_0} \tag{3.8}$$

where V_I is the incident voltage at the load, i.e. $V_I = V_i e^{-\gamma l}$, and V_R is the reflected voltage at the load, i.e. $V_R = \rho_v V_I$.
  At the load terminals $x = 0$ and the load voltage V_L is $V_L = V_I + V_R$. Also, $I_L Z_0 = V_I - V_R$, where I_L is the load current. Following the same steps as before leads to

$$V_x = V_L \cosh \gamma x + I_L Z_0 \sinh \gamma x \tag{3.9}$$

$$I_x = I_L \cosh \gamma x + \frac{V_L}{Z_0} \sinh \gamma x. \tag{3.10}$$

Equations (3.9) and (3.10) are the alternative form of the *general line equations*.

Voltage and Current Reflection Coefficients

The voltage reflection coefficient ρ_v of a mismatched line is the ratio (reflected voltage)/(incident voltage) at the load. Similarly, the current reflection coefficient ρ_i is the ratio (reflected current)/(incident current) at the load. Putting $x = 0$ in equations (3.7) and (3.8) makes it possible to obtain the load impedance in terms of the load current and voltage. Thus

$$Z_L = \frac{V_L}{I_L} = \frac{V_I + V_R}{\dfrac{V_I}{Z_0} - \dfrac{V_R}{Z_0}}.$$

Rearranging gives

$$\frac{Z_L}{Z_0} = \left[\frac{1 + V_R/V_I}{1 - V_R/V_I} \right] = \frac{1 + \rho_v}{1 - \rho_v}$$

Therefore,

$$\rho_v = \frac{Z_L - Z_0}{Z_L + Z_0}. \tag{3.11}$$

Since the reflected current is *always* in antiphase with the reflected voltage

$$\rho_i = -\rho_v = \frac{Z_0 - Z_L}{Z_0 + Z_L}.$$

There are two related terms that are sometimes referred to in the literature. These are:

(a) return loss $= 10 \log_{10} \left[\dfrac{\text{incident power}}{\text{reflected power}} \right]$ dB

$$= 20 \log_{10} \left[\frac{1}{\rho_v} \right] \text{dB}, \tag{3.12}$$

(b) reflection loss $= 10 \log_{10} \left[\dfrac{\text{incident power}}{\text{load power}} \right]$ dB

$$= 10 \log_{10} \left[\frac{1}{1 - \rho_v^2} \right] \text{dB}. \tag{3.13}$$

Example 3.2

An r.f. line has a characteristic impedance of 50 Ω, 3 dB loss, and it is $\lambda/2$ long. The line is terminated by a load of $100 + j20\ \Omega$. Calculate the voltage reflection coefficient at the load. The line is fed by a source of 50 Ω impedance and 2 V e.m.f. Calculate the sending-end and load voltages.

Solution
From equation (3.11),

$$\rho_v = \frac{100 + j20 - 50}{100 + j20 + 50} = 0.345 + j0.087 = 0.36 \angle 14°.$$

(*Ans.*)

The incident voltage is 1 V. 3 dB is a voltage ratio of $\sqrt{2}:1$ so that the incident voltage at the load is $0.707 \angle -180°$ V. The reflected voltage at the load is $0.707 \angle -180° \times 0.36 \angle 14° = 0.255 \angle -166°$ V. The reflected voltage at the sending end of the line is $0.255 \angle -166° \times 0.707 \angle -180°$ or $0.18 \angle 14°$ V.

(a) Sending-end voltage $= 1 + 0.18 \angle 14° = 1.175 + j0.044 = 1.176 \angle 2.2°$ V. (*Ans.*)

(b) Load voltage $= 0.707 \angle -180° + 0.255 \angle -166° = 0.956 \angle -176°$ V. (*Ans.*)

Voltage Reflection Coefficient at any Point on a Line

At the mismatched load $\rho_v = V_I/V_R$. At any distance x from the load

$$\rho_{v(x)} = \frac{V_R e^{-\gamma x}}{V_I e^{\gamma x}}$$

or $\rho_{v(x)} = \rho_v e^{-2\gamma x}$. (3.14)

Example 3.3

A loss-free line has a characteristic impedance of 60 Ω and a load impedance of 120 Ω. Calculate the voltage reflection coefficient (*a*) at the load, (*b*) at a distance of $\lambda/8$ from the load, and (*c*) $\lambda/8$ from the load if the line loss is 8 dB per wavelength.

Solution

(a) $\rho_v = \dfrac{120 - 60}{120 + 60} = 1/3$. (*Ans.*)

(b) $\rho_{v(x)} = \frac{1}{3} e^{-2(0+j\pi/4)} = \frac{1}{3} e^{-j\pi/2} = \frac{1}{3} \angle -90°$. (*Ans.*)

(c) $\rho_{v(x)} = \frac{1}{3} e^{-2(0.115+j\pi/4)} = \frac{1}{3} e^{-0.223} \angle -90° = 0.267 \angle -90°$. (*Ans.*)

Input Impedance of a Mismatched Line

The impedance at any point along a mismatched line is the ratio of the total voltage to the total current at that point. The input impedance of a line is the ratio (sending-end voltage)/(sending-end current). Using equations (3.2) and (3.4) with $x = 0$

$$Z_S = \frac{V_S}{I_S} = \frac{V_i + V_r}{\dfrac{V_i}{Z_0} - \dfrac{V_r}{Z_0}} = Z_0 \left[\frac{V_i + V_r}{V_i - V_r} \right].$$

Now, $V_r = \rho_{v(l)} V_I = \rho_v V_I e^{-\gamma l} = \rho_v e^{-2\gamma l} V_i$ so that

$$Z_S = Z_0 \left[\frac{1 + \rho_v e^{-2\gamma l}}{1 - \rho_v e^{-2\gamma l}} \right].$$ (3.15)

Alternatively, from equations (3.9) and (3.10),

$$Z_S = \frac{V_L \cosh \gamma l + I_L Z_0 \sinh \gamma l}{I_L \cosh \gamma l + \dfrac{V_L}{Z_0} \sinh \gamma l}$$

or $Z_S = Z_0 \left[\dfrac{Z_L \cosh \gamma l + Z_0 \sinh \gamma l}{Z_0 \cosh \gamma l + Z_L \sinh \gamma l} \right].$ (3.16)

If the load terminals are short-circuited so that $Z_L = 0$, then

$$Z_S = Z_0 \tanh \gamma l. \tag{3.17}$$

Similarly, for an open-circuited line $Z_L = \infty$ and $Z_S = Z_0 \cosh \gamma l$.

Example 3.4

A line is $3\lambda/2$ long and has a characteristic impedance of 50 Ω and 3 dB loss. Calculate its input impedance when the load impedance is 100 Ω.

Solution

(a) $\rho_v = \dfrac{100 - 50}{100 + 50} = \frac{1}{3}$.   3 dB = 0.345 nepers.

Substituting into equation (3.15)

$$Z_S = 50 \left[\frac{1 + \frac{1}{3} e^{-2(0.345 + j3\pi)}}{1 - \frac{1}{3} e^{-2(0.345 + j3\pi)}} \right] = 50 \left[\frac{1 + \frac{1}{3} e^{-0.69}}{1 - \frac{1}{3} e^{0.69}} \right] = 70\ \Omega.$$

(Ans.)

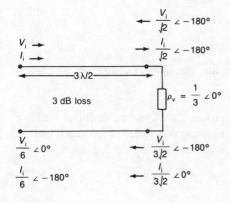

Fig. 3.2

(b) Work from first principles, see Fig. 3.2. The total sending-end voltage is

$$V_S = V_i + \frac{V_i}{6} = \frac{7}{6} V_i$$

and the total sending-end current is

$$I_S = I_i - \frac{I_i}{6} = \frac{5}{6} I_i.$$

Therefore,

$$Z_S = \frac{V_S}{I_S} = \frac{7 V_i}{6} \times \frac{6}{5 I_i} = \frac{7}{5} Z_0 = 70\ \Omega. \quad \text{(Ans.)}$$

Low-loss Lines

A low-loss line is one whose attenuation αl is small enough for the approximations $\cosh \alpha l \simeq 1$ and $\sinh \alpha l \simeq \alpha l$ to be valid. The input impedance of a short-circuited line, see equation (3.17), can then be written as

$$Z_S = Z_0 \left[\frac{\sinh (\alpha + j\beta) l}{\cosh (\alpha + j\beta) l} \right]$$

$$= Z_0 \left[\frac{\sinh \alpha l \cosh j\beta l + \sinh j\beta l \cosh \alpha l}{\cosh \alpha l \cosh j\beta l + \sinh \alpha l \sinh j\beta l} \right]$$

$$= Z_0 \left[\frac{\sinh \alpha l \cos \beta l + j\cosh \alpha l \sin \beta l}{\cosh \alpha l \cos \beta l + j \sinh \alpha l \sin \beta l} \right].$$

(Sinh $jx = j \sin x$ and $\cosh jx = \cos x$.)

$$Z_S \simeq Z_0 \left[\frac{\alpha l \cos \beta l + j \sin \beta l}{\cos \beta l + j \sin \beta l \alpha l} \right]. \tag{3.18}$$

(a) When $l = \lambda/4$, then $\cos \beta l = 0$ and $\sin \beta l = 1$. Then

$$Z_S = Z_0 \left[\frac{j}{j \alpha l} \right] = \frac{Z_0}{\alpha l}. \tag{3.19}$$

(b) When $l = \lambda/2$, then $\cos \beta l = -1$ and $\sin \beta l = 0$. Now

$$Z_S = Z_0 \left[\frac{-\alpha l}{-1} \right] = Z_0 \alpha l. \tag{3.20}$$

Loss-free Lines

Very often the attenuation of an r.f. line is small enough to be neglected so that $\alpha l \simeq 0$ and $\gamma l \simeq j\beta l$. Then equation (3.16) becomes

$$Z_S = Z_0 \left[\frac{Z_L \cos \beta l + j Z_0 \sin \beta l}{Z_0 \cos \beta l + j Z_L \sin \beta l} \right]. \tag{3.21}$$

Short-circuited Line

If the load terminals of the line are short-circuited

$$Z_L = 0 \text{ and } Z_S = j Z_0 \tan \beta l. \tag{3.22}$$

This means that the input impedance of a loss-free short-circuited line is a pure reactance whose magnitude and sign are determined by both the characteristic impedance of the line and the line's length.

Open-circuited Line

If the load terminals of a line are left open-circuit, $Z_L = \infty$ and

$$Z_S = -j Z_0 \cos \beta l. \tag{3.23}$$

$\lambda/4$ Length of Line

When the electrical length of a loss-free line is exactly one-quarter of a wavelength $\gamma l = j\beta l = j\pi/2$. Then $\cos \beta l = 0$ and $j \sin \beta l = j$ so that equation (3.21) becomes

$$Z_S = \frac{Z_0^2}{Z_L}. \tag{3.24}$$

$\lambda/2$ Length of Line

For a $\lambda/2$ length of loss-free line $\gamma l = j\beta l = j\pi$ and so $\cos \beta l = -1$ and $\sin \beta l = 0$, giving an input impedance of

$$Z_S = Z_0 \left[\frac{-Z_L}{-Z_0} \right] = Z_L. \tag{3.25}$$

This means that the input impedance of a $\lambda/2$ length of loss-free line is equal to the load impedance.

$\lambda/8$ Length of Line

Now $\beta l = \pi/4$ and $\cos \beta l = \sin \beta l = 1/\sqrt{2}$. Hence

$$Z_S = Z_0 \left[\frac{\dfrac{Z_L}{\sqrt{2}} + \dfrac{jZ_0}{\sqrt{2}}}{\dfrac{Z_0}{\sqrt{2}} + \dfrac{jZ_L}{\sqrt{2}}} \right] = Z_0 \underline{\bigg/ \tan^{-1}\left(\frac{Z_0}{Z_L} - \frac{Z_L}{Z_0} \right)}. \quad (3.26)$$

This means that the magnitude of the input impedance of a $\lambda/8$ length of loss-free line is equal to the characteristic impedance of the line.

Example 3.5

Calculate the impedance Z_0^1 of the $\lambda/4$ section of line shown in Fig. 3.3, if the input impedance Z_S is equal to 50 Ω.

Solution

Since the length of line from the load to the point B is $\lambda/2$ the input impedance at B is 200 Ω and the total load impedance for the $\lambda/4$ section is 100 Ω. The impedance required at the point A, for the input impedance of the system to be 50 Ω, is also 50 Ω. Hence,

$$Z_0^1 = \sqrt{(50 \times 100)} = 70.7 \ \Omega. \quad (Ans.)$$

Standing Waves and Voltage Standing Wave Ratio

At any point along a mismatched line the total voltage, or current, is the phasor sum of the incident and the reflected waves at that point. If the r.m.s. values of the total current and voltage are plotted against distance from the load, *standing waves* will be obtained.

Consider a loss-free line. The maximum voltage V_{max} on the line will occur whenever the incident and reflected waves are in phase with one another, i.e. $V_{max} = V_i + V_r = V_i(1 + |\rho_v|)$. The minimum voltage V_{min} occurs at those points on the line where the incident and the reflected waves are in antiphase with one another, i.e. $V_{min} = V_i - V_r = V_i(1 - |\rho_v|)$. The points at which the

Fig. 3.3

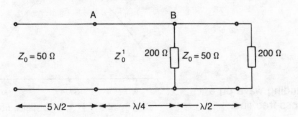

maximum voltage occurs are known as antinodes, whilst the point of minimum voltage are known as nodes. Figure 3.4 shows the standing wave of voltage on a line of characteristic impedance 600 Ω and load impedance 200 Ω. The voltage reflection coefficient is equal to $(200 - 600)/(200 + 600) = -0.5$ and so the maximum and minimum voltages on the line are $1.5 V_i$ and $0.5 V_i$, respectively. The minimum voltage occurs at the load and at multiples of $\lambda/2$ from the load. The maximum voltage occurs $\lambda/4$ from the load and then at multiples of $\lambda/2$ from that point.

The *voltage standing wave ratio S* is the ratio (maximum voltage)/ (minimum voltage) or the ratio (minimum voltage)/(maximum voltage). Either definition can be used as required; no confusion should result since the v.s.w.r. will always be *either* greater than, or less than, unity. Using the first definition

$$S = \frac{V_{max}}{V_{min}} = \frac{V_i(1 + |\rho_v|)}{V_i(1 - |\rho_v|)} = \frac{1 + |\rho_v|}{1 - |\rho_v|}. \quad (3.27)$$

For the line whose standing wave is shown by Fig. 3.4, $|\rho_v| = 0.5$ and $S = 3$.

There are a number of reasons why the presence of standing waves on a line used to transmit energy from one point to another is undesirable. These reasons are as follows.

(*a*) If the load is not matched to the line the maximum transfer of power to the load will not take place. The incident power is V_I^2/Z_0 and the reflected power is V_R^2/Z_0. The power dissipated in the load is

$$P_L = \frac{V_I^2 - V_R^2}{Z_0} = \frac{(V_I + V_R)(V_I - V_R)}{Z_0} = \frac{V_{max} V_{min}}{Z_0}$$

or $\quad P_L = \dfrac{V_{max}^2}{S Z_0}$. $\quad (3.28)$

(*b*) The reflected current and voltage waves are attenuated as they

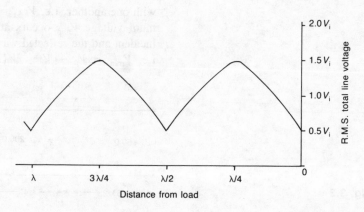

Fig. 3.4 Standing wave on a mismatched loss-free line.

Distance from load

travel back towards the sending end of the line. This means that the total line loss is increased.

(c) At an antinode the voltage is $V_{\max} = V_i(1 + |\rho_v|)$ and it may be anything up to twice as great as the incident voltage. Since the breakdown voltage of the dielectric between the conductors of a line must not be exceeded, this limits the maximum possible peak value of the incident voltage, and hence the incident power which the line can transmit.

v.s.w.r. on a Lossy Line

When the loss of a line is not negligibly small the incident and reflected waves are attenuated as they travel. As the distance from the load increases the incident voltage will get bigger and the reflected voltage will get smaller, and so the v.s.w.r. will become smaller, see Fig. 3.5. The voltage reflection coefficient at distance x from the load is $\rho_{v(x)} = \rho_v e^{-2\gamma x}$, and therefore the v.s.w.r. at this point is

$$S_{(x)} = \frac{1 + |\rho_v e^{-2\gamma x}|}{1 - |\rho_v e^{-2\gamma x}|} = \frac{1 + |\rho_v|e^{-2\alpha x}}{1 - |\rho_v|e^{-2\alpha x}}. \tag{3.29}$$

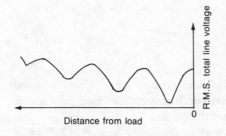

R.M.S. total line voltage

Distance from load

Fig. 3.5 Standing wave on a mismatched lossy line.

Example 3.6

A line has an attenuation of 6 dB and the v.s.w.r. near to the load is 2. Calculate the v.s.w.r. near the sending-end of the line.

Solution
From equation (3.29)

$$S_{(x)} = \frac{1 + \left(\dfrac{S - 1}{S + 1}\right)e^{-2\alpha x}}{1 - \left(\dfrac{S - 1}{S + 1}\right)e^{-2\alpha x}} = \frac{(S + 1)e^{\alpha x} + (S - 1)e^{-\alpha x}}{(S + 1)e^{\alpha x} - (S - 1)e^{-\alpha x}}$$

$\alpha x = 6/8.686 = 0.69$ nepers, so $e^{\alpha x} \simeq 2$ and $e^{-\alpha x} = 0.5$. Hence

$$S_{(x)} = \frac{3 \times 2 + 1 \times 0.5}{3 \times 2 - 1 \times 0.5} = \frac{6.5}{5.5} = 1.18. \quad (Ans.)$$

Line Mismatched at Both Ends

If the impedance of the source connected to the sending-end terminals of the line is also not equal to the characteristic impedance of the line reflections will occur at both ends of the line. The voltage and current waves reflected at the load will travel back to the sending end of the line. Here some of the reflected energy will be reflected again to travel over the line towards the load. Another reflection at the load takes place and the reflected energy is transmitted back towards the source. This will happen repeatedly until the effect of line attenuation reduces the reflected waves to a negligibly small value. The situation on a

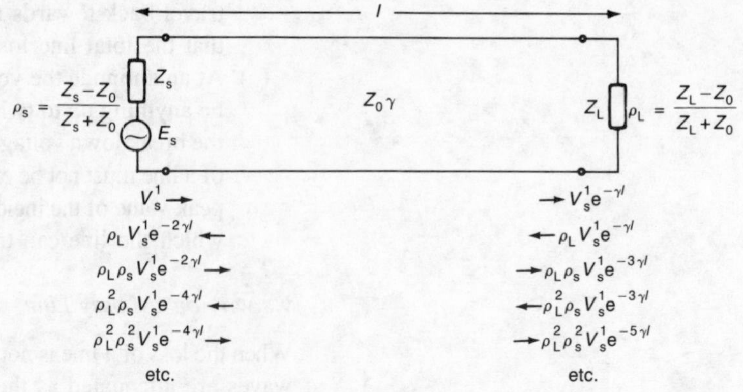

$$\rho_s = \frac{Z_s - Z_0}{Z_s + Z_0}$$

$$Z_0 \gamma$$

$$\rho_L = \frac{Z_L - Z_0}{Z_L + Z_0}$$

$V_s^1 \rightarrow$ $\rightarrow V_s^1 e^{-\gamma l}$

$\rho_L V_s^1 e^{-2\gamma l} \leftarrow$ $\leftarrow \rho_L V_s^1 e^{-\gamma l}$

$\rho_L \rho_s V_s^1 e^{-2\gamma l} \rightarrow$ $\rightarrow \rho_L \rho_s V_s^1 e^{-3\gamma l}$

$\rho_L^2 \rho_s V_s^1 e^{-4\gamma l} \leftarrow$ $\leftarrow \rho_L^2 \rho_s V_s^1 e^{-3\gamma l}$

$\rho_L^2 \rho_s^2 V_s^1 e^{-4\gamma l} \rightarrow$ $\rightarrow \rho_L^2 \rho_s^2 V_s^1 e^{-5\gamma l}$

Fig. 3.6 Line mismatched at both ends.

etc.   etc.

line mismatched at both ends is illustrated by Fig. 3.6, where ρ_L and ρ_S are the voltage reflection coefficients at the load, and at the sending-end terminals, respectively, and $V_S^1 = (E_S Z_0)/(Z_S + Z_0)$. The total voltage at any point along the line is the sum to infinity of the incident and reflected voltages at that point. The voltages at any point have values that form a geometric progression (or series), and thus their sum to infinity is

$$S = \frac{\text{initial term}}{1 - \text{common ratio}}.$$

Sending-end Voltage

Source-to-load direction
Initial term $= V_S^1$, common ratio $= \rho_L \rho_S e^{-2\gamma l}$. Hence

$$V_1 = \frac{V_S^1}{1 - \rho_L \rho_S e^{-2\gamma l}}.$$

Load-to-source direction
Initial term $= \rho_L V_S^1 e^{-2\gamma l}$, common ratio $= \rho_L \rho_S e^{-2\gamma l}$. Hence

$$V_2 = \frac{\rho_L V_S^1 e^{-2\gamma l}}{1 - \rho_L \rho_S e^{-\gamma l}}.$$

The sending-end voltage is the *sum* of these voltages, i.e.

$$V_S = \frac{V_S^1 (1 + \rho_L e^{-2\gamma l})}{1 - \rho_L \rho_S e^{-2\gamma l}}. \tag{3.30}$$

Load Voltage

Source-to-load direction
Initial value $= V_S^1 e^{-\gamma l}$, common ratio $= \rho_L \rho_S e^{-2\gamma l}$. Hence

$$V_1 = \frac{V_S^1 e^{-\gamma l}}{1 - \rho_L \rho_S e^{-2\gamma l}}.$$

Load-to-source direction

Initial value $= \rho_L V_S^1 e^{-\gamma l}$, common ratio $= \rho_L \rho_S e^{-2\gamma l}$. Hence

$$V_2 = \frac{\rho_L V_S^1 e^{-\gamma l}}{1 - \rho_L \rho_S e^{-2\gamma l}}.$$

The total load voltage V_L is the sum of V_1 and V_2, or

$$V_L = \frac{V_S^1 e^{-\gamma l}(1 + \rho_L)}{1 - \rho_L \rho_S e^{-2\gamma l}}. \tag{3.31}$$

Example 3.7

A line of characteristic impedance 600 Ω and 3 dB loss is $\lambda/2$ in length. It is fed by a source of e.m.f. 5 V and impedance 1000 Ω and it is terminated by a load of 400 Ω impedance. Calculate the load voltage.

Solution

$$\rho_S = \frac{1000 - 600}{1000 + 600} = 0.25; \; \rho_L = \frac{400 - 600}{400 + 600} = -0.25.$$

$$V_S^1 = \frac{5 \times 600}{600 + 1000} = 1.875 \text{ V}.$$

$$e^{-\gamma l} = e^{-(\alpha + j\beta)l} = 0.707 \; \angle -180° = -0.707.$$

$$e^{-2\gamma l} = e^{-2(\alpha + j\beta)l} = 0.5 \; \angle -360° = 0.5.$$

Therefore,

$$V_L = \frac{1.875 \times -0.707 \times 0.75}{1 - (-0.25 \times 0.25 \times 0.5)} = -0.964 \text{ V}. \quad (Ans.)$$

Transmission Lines as Components

At the higher end of the v.h.f. band, and above, the use of discrete components such as capacitors and, particularly, inductors becomes increasingly difficult, and very often a transmission line is used to simulate a wanted component. The input impedance Z_S of a loss-free line short-circuited at its load terminals is given by $Z_S = jZ_0 \tan \beta l$. Such a line has an input impedance that is very nearly a pure reactance; if the electrical length βl of the line is less than $\lambda/4$ the line will simulate an inductor; if $\lambda/4 < \beta l < \lambda/2$ a capacitor will be simulated. Longer lengths are not employed except at u.h.f. because the line losses will no longer be negligibly small. If the line length is approximately equal to $\lambda/4$, or to $\lambda/2$, its impedance will vary with frequency in a similar manner to that of a series-, or parallel-tuned circuit.

Suppose, for example, that an inductance of 100 nH is wanted at a frequency of 600 MHz and that it is to be simulated by a short-circuited line of characteristic impedance 50 Ω. The reactance required is $j2\pi \times 600 \times 10^6 \times 100 \times 10^{-9} = j377 \; \Omega$, so that $377 = 50 \tan \beta l$ or $\beta l = \tan^{-1} 7.54 = 1.439$ rad. Hence, $2\pi l/\lambda = 1.439$

or $l = 1.439\lambda/2\pi = 0.229\lambda$. The wavelength λ is 0.5 m so that the physical length of the short-circuited line should be $0.229 \times 0.5 = 0.1145$ m $= 11.45$ cm.

A similar calculation can be made to determine the length of line needed to simulate a capacitive reactance. If the length thus calculated is inconveniently long an open-circuited line will give a shorter length but there may well be radiation problems from the end of the line. When the electrical length of the line is very short, certainly less than $\lambda/8$, $\tan \beta l \simeq \beta l$ and so the simulated inductive reactance $X_L \simeq Z_0\beta l = \sqrt{L/C} \times \omega \sqrt{LC} \times l = \omega Ll$. Then the simulated inductance is given by Ll where L is the inductance per metre and l is the length of the line.

The Q-factor of the simulated inductance is

$$Q = \frac{\omega L}{r} = \frac{\omega L\dagger}{2\alpha\sqrt{L/C}} = \frac{\sqrt{LC} \cdot \omega}{2\alpha} = \frac{\beta}{2\alpha} = \frac{2\pi}{2\lambda\alpha},$$

or $Q = \dfrac{\pi}{\alpha\lambda}$. \hfill (3.32)

Clearly, the Q-factor will increase with both frequency and with a reduction in the attenuation of the line. If the attenuation coefficient of the line is 8×10^{-3} nepers per metre and λ is 0.5 m then $Q = 785$.

A Line as a Tuned Circuit

A length of loss-free line can also be employed to simulate either a series-, or parallel-tuned circuit. The possible simulations are shown in Table 3.1.

Table 3.1

Electrical length	Short-circuited	Open-circuited
$\lambda/4$	Parallel-tuned	Series-tuned
$\lambda/2$	Series-tuned	Parallel-tuned

Consider a low-loss short-circuited line whose input impedance is

$$Z_S = Z_0\left[\frac{\alpha l \cos \beta l + j \sin \beta l}{\cos \beta l + j \sin \beta l\alpha l}\right]$$

(equation 3.18) again).

$\dagger$From $\alpha = R/2Z_0$.

(a) For $1 \simeq \lambda/4 \; \beta l = \pi/4$, so that $\cos \beta l$ is small and $\sin \beta l \simeq 1$. Therefore,

$$Z_S \simeq Z_0 \left[\frac{j}{\cos \beta l + j\alpha l} \right] = \frac{Z_0}{\alpha l} \left[\frac{1}{1 - \dfrac{j \cos \beta l}{\alpha l}} \right].$$

Now

$$\cos \beta l = \sin (90° - \beta l) \simeq 90° - \beta l = \beta_0 l - \beta l$$

$$= \beta_0 \left[1 - \frac{\beta_0}{\beta} \right] l$$

and $\beta = \omega/v_p$ so

$$\cos \beta l = \beta_0 \left[1 - \frac{\omega}{\omega_0} \right] l = \beta_0 \left[\frac{\omega_0 - \omega}{\omega_0} \right] l.$$

Therefore,

$$Z_S = \frac{Z_0}{\alpha l} \left[\frac{1}{1 - \left[j2\beta_0 \left(\dfrac{\omega_0 - \omega}{\omega_0} \right) l \right] \Big/ 2\alpha l} \right]. \qquad (3.33)$$

The impedance of a parallel-tuned circuit can be written in the form

$$Z = \frac{R_d}{1 + jQ2 \left(\dfrac{\omega_0 - \omega}{\omega_0} \right)}$$

(see *Electrical Principles IV*) and it is evident that the two impedances vary with frequency in a similar manner. Comparing the two equations

$$Q = \frac{\beta_0}{2\alpha} = \frac{2\pi}{2\lambda\alpha} = \frac{\pi}{\alpha\lambda}.$$

(b) For $l \simeq \lambda/2$, $\beta l \simeq \pi$, $\cos \beta l = -1$ and $\sin l$ is small. Hence

$$Z_S = Z_0 \left[\frac{-\alpha l + j \sin \beta l}{-1} \right] = Z_0 \alpha l \left[1 - \frac{j \sin \beta l}{\alpha l} \right].$$

Now

$$\sin \beta l = -\sin (-\beta l) = -\sin (-\pi + \beta l)$$

$$= -\sin (-\beta_0 l + \beta l) \simeq \beta_0 l - \beta l$$

$$= \beta_0 \left[1 - \frac{\omega}{\omega_0} \right] l.$$

Hence

$$Z_S = Z_0 \alpha l \left[\frac{1 - \beta_0 \left(\dfrac{\omega_0 - \omega}{\omega_0} \right)}{\alpha} \right],$$

which is of the same form as the impedance of a series-tuned circuit, i.e.

$$Z = R \left[1 + jQ \left(\frac{\omega_0 - \omega}{\omega_0} \right) \right].$$

Matching

For the maximum transfer of energy from a line to its load the impedance of the load must be equal to the characteristic impedance of the line. Wherever possible the load impedance is selected to satisfy this requirement. Very often, however, the two impedances cannot be made equal to one another and then some form of matching device may be used. The majority of line matching systems are of one or another of the following forms: (*a*) λ/4 low-loss matching sections, (*b*) single, or double stubs, or (*c*) baluns (see *Radio Systems for Technicians*).

λ/4 Matching Sections

A quarter-wavelength (λ/4) length of loss-free line has an input impedance Z_S given by equation (3.24), i.e. $Z_S = Z_0^2/Z_L$. This means that a load impedance Z_L can be transformed into a desired value of input impedance Z_S by the suitable choice of the characteristic impedance Z_0. The required value of Z_0 is easily obtained by transposing equation (3.24) to give

$$Z_0 = \sqrt{Z_L Z_S}. \tag{3.34}$$

If the impedance of the load is purely resistive the λ/4 matching section can be connected between the line and the load as shown by Fig. 3.7. If, however, the load has a reactive component the λ/4 section must be inserted into the line at a distance from the load at which the impedance of the line is purely resistive. This is a calculation that is best performed with the aid of a Smith chart (p. 80).

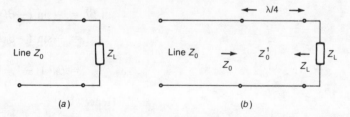

Fig. 3.7 Use of a λ/4 matching section.

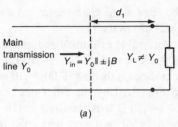

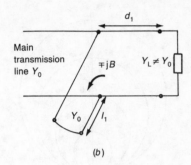

Fig. 3.8 (a) Single-stub matching,
(b) double-stub matching.

Stub Matching

The admittance of a mismatched line varies with the distance from the load. At some particular distance d_1 the admittance will be equal to the *characteristic admittance* $Y_0 = 1/Z_0$ of the line, in parallel with a susceptance B. At this point $Y_{in} = Y_0 \pm jB$, Fig. 3.8(a). If this susceptance can be cancelled out by another susceptance of equal magnitude but of the opposite sign, the input admittance of the line will become $Y_{in} = Y_0$. The line will then be matched at this point.

The necessary susceptance is provided by the connection of a short-circuited *stub line* in parallel with the line at distance d_1 from the load. This is shown by Fig. 3.8(b). Not all values of load impedance can be matched in this way and it will sometimes be necessary to employ two stubs, as shown by Fig. 3.18. Stub matching problems are also best tackled with the aid of a Smith chart.

The Smith Chart

The Smith chart is a plot of normalized impedance against the magnitude and angle of voltage reflection coefficient. It can be used in the solution of many problems involving transmission lines (and waveguides) since it can often greatly simplify a problem. The Smith chart consists of: (a) a real axis with values which vary from zero to infinity, with unity in the centre; (b) a series of circles centred on the real axis; and (c) a series of arcs of circles that start from the infinity point on the real axis. This is shown by Figs 3.9(a) and (b). The circles represent the *real* parts of the normalized impedances, i.e. R/Z_0, and the arcs represent the imaginary parts of the normalized impedances, i.e. $\pm jX/Z_0$.

Figure 3.10 shows a full Smith chart. In addition to the circles and the arcs of circles representing normalized resistance and reactance the edge of the chart is marked with scales of (a) angle of reflection coefficient (in degrees), and (b) distance (in wavelengths). Movement around the edge of the chart in the *clockwise* direction corresponds to movement along the line towards the source; conversely, anti-clockwise movement around the chart represents movement along the line towards the load. It should be noticed that a complete circle around

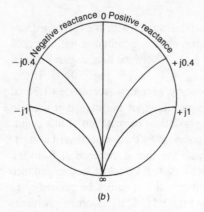

(a)

(b)

Fig. 3.9 (a) Real parts, and (b) imaginary parts of normalized impedance Z/Z_0.

the edge of the chart represents a movement of one-half wavelength ($\lambda/2$) along the line.

Any value of normalized impedance can be located on the chart. Suppose that $Z = 100 + j100\ \Omega$ and $Z_0 = 50\ \Omega$. The normalized impedance is then $z = (100 + j100)/50 = 2 + j2$; this is shown plotted in Fig. 3.10 by the point marked as A. Similarly, an impedance $10 - j20\ \Omega$ is normalized to $(10 - j20)/50 = 0.2 - j0.4$ and is represented on the chart by the point B. Admittances can also be plotted on the Smith chart; the admittance must first be normalized by dividing it by the characteristic admittance Y_0 of the line. Thus, if $Y = 0.02 - j0.03\ S$ and $Z_0 = 50\ \Omega$ then $y = (0.02 - j0.03)/0.02 = 1 - j1.5$ and is plotted on the chart as the point C.

Use of the Smith Chart

Essentially, the Smith chart deals with lines of negligible loss; the effect of any line attenuation can be taken into account by the use of a separate scale and this will be dealt with later.

Voltage Reflection Coefficient

To determine the voltage reflection coefficient produced by a load impedance the impedance must first be normalized and located on the chart. A straight line should then be drawn from the centre of the chart (the point $1 + j0$), through the plotted point z to the edge of the chart. The distance of the point from the centre divided by the distance centre-to-edge is equal to the magnitude of the voltage reflection coefficient. The phase angle of the voltage reflection coefficient is read from the scale at the edge of the chart.

Example 3.8

Calculate the voltage reflection coefficient of the line in Example 3.2.

Solution

The load impedance $Z_L = 100 + j20\ \Omega$ and the characteristic impedance $Z_0 = 50\ \Omega$ so that $z_L = 2 + j0.4$. This point is plotted on the Smith chart as shown by Fig. 3.11. The line drawn from the point $1 + j0$ through z_L passes through the voltage reflection coefficient angle scale at 14°. The distance from the point $(1 + j0)$ to z_L is 30.6 mm and the distance from $(1 + j0)$ to the edge of the chart is 85 mm and so

$$|\rho_v| = 30.6/85 = 0.36.$$

Therefore, voltage reflection coefficient $= 0.36\ \angle 14°$.   (*Ans.*)

The procedure is slightly different when a load admittance is involved. First, locate the normalized admittance y_L on the chart and then draw a straight line from y_L, through the centre of the chart, to the edge

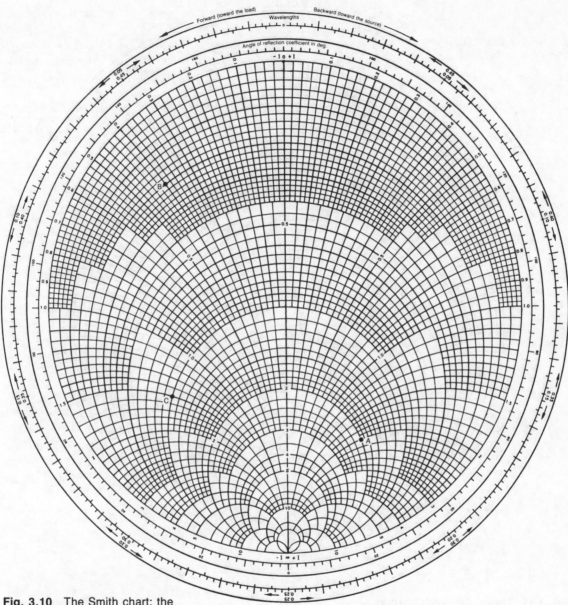

Fig. 3.10 The Smith chart; the point A represents $z = 2 + j2$, the point B represents $z = 0.2 - j0.4$ and point C represents $y = 1 - j1.5$.

of the chart. Then the magnitude of the voltage reflection coefficient $|\rho_v|$ is equal to the ratio (distance y_L to centre)/(distance edge to centre), and the angle of y_L is read from the scale at the edge of the chart.

Voltage Standing-wave Ratio

To calculate the v.s.w.r. on a line locate the normalized impedance z_L (or the normalized admittance y_L) on the chart and draw a circle,

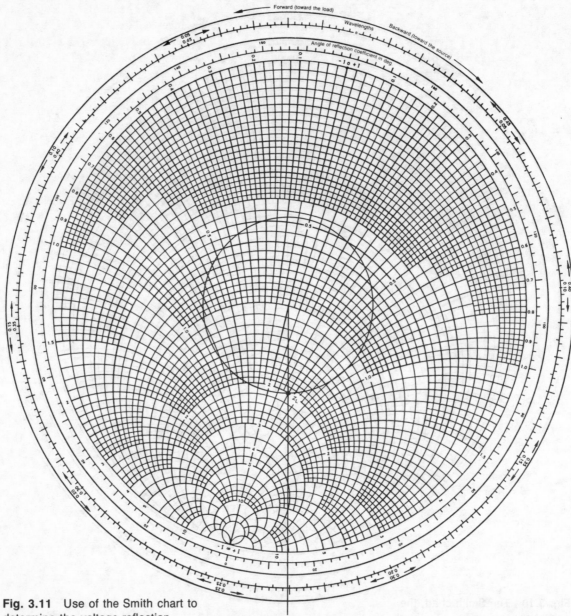

Fig. 3.11 Use of the Smith chart to determine the voltage reflection coefficient.

centred on the point $(1 + j0)$, which passes through the point z_L. The v.s.w.r. is then equal to the value of the real axis at the point where it is cut by the circle. Two values will be obtained: one greater than, and the other less than, unity and each will be the reciprocal of the other. Referring to Fig. 3.11 the v.s.w.r. is about 2.1 or 0.47.

Input Impedance of a Length of Line

To determine the input impedance of a length of loss-free line locate the normalized load impedance on the chart and draw an arc of a circle, centred on the centre of the chart, i.e. the point $(1 + j0)$, moving clockwise. The length of the arc, measured on the wavelength scale, should be equal to the electrical length of the line. The normalized input impedance of the line is then given by the location of the end of the arc. The procedure can be reversed if the input impedance is known and the load impedance is to be determined.

Example 3.9

A 50 Ω line has a load impedance of $20 - j20$ Ω. Use the Smith chart to find (*a*) the voltage reflection coefficient, (*b*) the v.s.w.r., (*c*) the input impedance of a 0.2λ length of this line, and (*d*) the lengths of line that have a purely resistive input impedance and the values of these resistances.

Solution
The normalized load impedance is

$$z_L = \frac{20 - j20}{50} = 0.4 - j0.4$$

and this is plotted on the chart shown in Fig. 3.12.

(*a*) $|\rho_v| = \dfrac{42.5 \text{ mm}}{85 \text{ mm}} = 0.5, \; \angle\rho_v = -131°.$

Therefore, $\rho_v = 0.5 \angle -131°.$ (*Ans.*)
(*b*) $S = 3$ or 0.33.
(*c*) Travelling around the $S = 3$ circle a distance of 0.2λ from z_L towards the source gives $z_{in} = 0.65 + j0.85$. Therefore,

$$Z_{in} = (0.65 + j0.85)50 = 32.5 + j42.5 \; \Omega. \quad (Ans.)$$

(*d*) Travelling from z_L towards the source 0.068λ gives $z_{in} = 0.33 + j0$, and hence $Z_{in} = 16.67$ Ω. (*Ans.*)
     Travelling from z_L towards the source $(0.068 + 0.25)\lambda = 0.318\lambda$ makes $z_{in} = 3 + j0$ and $Z_{in} = 150$ Ω. (*Ans.*)

Simulation of a Component

When a length of loss-free line is used to simulate an inductance or a capacitance it is nearly always short-circuited at the load terminals. Then $Z_L = z_L = 0$ and the input reactance of the line can be found by moving clockwise around the outside of the chart.

Example 3.10

A length of loss-free short-circuited line has $Z_0 = 50$ Ω and is to simulate (*a*) an inductive reactance, and (*b*) a capacitive reactance, of 35 Ω at 600 MHz. Calculate the necessary lengths of line.

Solution
(*a*) $x_{in} = j35/50 = j0.7$. From the Smith chart this is a distance of 0.0962λ from the top of the real axis. At 600 MHz $\lambda = 0.5$ m and so the length needed is 4.81 cm. (*Ans.*)

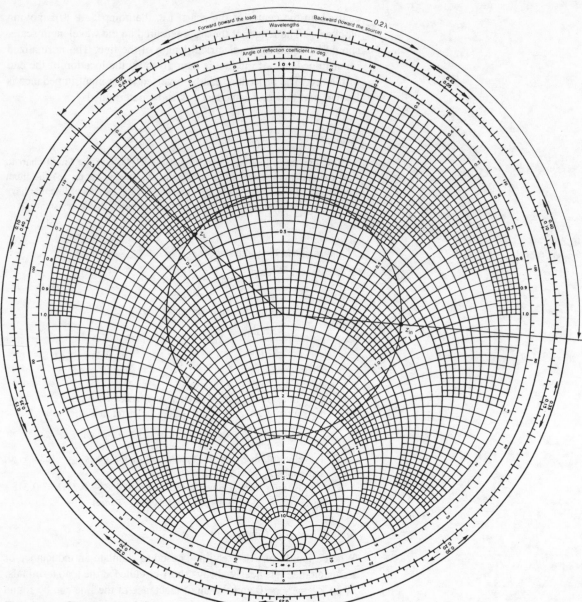

Fig. 3.12 Calculation of the v.s.w.r. and input impedance of a line.

(b) $x_{in} = -j35/50 = -j0.7$. Now the length needed is 0.3063λ or 15.32 cm.   (*Ans.*)

Determination of an Unknown Impedance

If both the length of the line and its input impedance are known the method previously described can be used. If not, the procedure to be adopted is as follows.

(a) Measure the v.s.w.r. with the unknown load connected to the line and note the position of *any* voltage minimum.

(*b*) Remove the load from the line and short-circuit the load terminals. This will cause the noted position of the voltage minimum to shift to a new point (which is less than λ/4 away). Note this new position.

(*c*) Measure the distance in centimetres between two adjacent voltage minima — this corresponds to one-half a wavelength on the line.

(*d*) Draw the v.s.w.r. circle.

(*e*) Starting from the point where the v.s.w.r. circle cuts the real axis at a value less than unity, move around the v.s.w.r. circle a distance equal to the distance moved by the voltage minimum in (*b*) and in the *same* direction. The point reached is the normalized load impedance.

If an unknown load admittance is to be determined follow the same procedure but start on the Smith chart at the point where the v.s.w.r. circle cuts the real axis at a value greater than unity.

Example 3.11

A 50 Ω line has a v.s.w.r. of 2 when an unknown load impedance is connected to its output terminals. Adjacent voltage minima are found to be 30 cm apart. When the unknown load is removed from the line and is replaced by a short-circuit the voltage minima moves by 7.5 cm towards the source. Calculate the value of the unknown load impedance.

Solution
See Fig. 3.13. The v.s.w.r. = 2 circle has been drawn. 30 cm = λ/2 so that 7.5 cm = 0.125λ. Moving around the v.s.w.r. circle for this distance in the clockwise direction gives $z_L = 0.8 + j0.58$. Therefore

$$Z_L = 50(0.8 + j0.58) = 40 + j29 \ \Omega. \quad (Ans.)$$

Effect of Line Attenuation

The effect of the line attenuation upon the v.s.w.r. and the input impedance can be taken into account by the use of the scales shown in Fig. 3.14. The line attenuation scale is marked in 1 dB steps and may be entered at any point since only distances along the scale are of any significance. The use of the scales will be illustrated by an example.

Example 3.12

A 50 Ω line has 3 dB loss and is terminated in a 200 + j25 Ω load. Calculate its input resistance if the line is 0.3λ long.

Solution
Refer to Fig. 3.15;

$$z_L = \frac{200 + j25}{50} = 4 + j0.5.$$

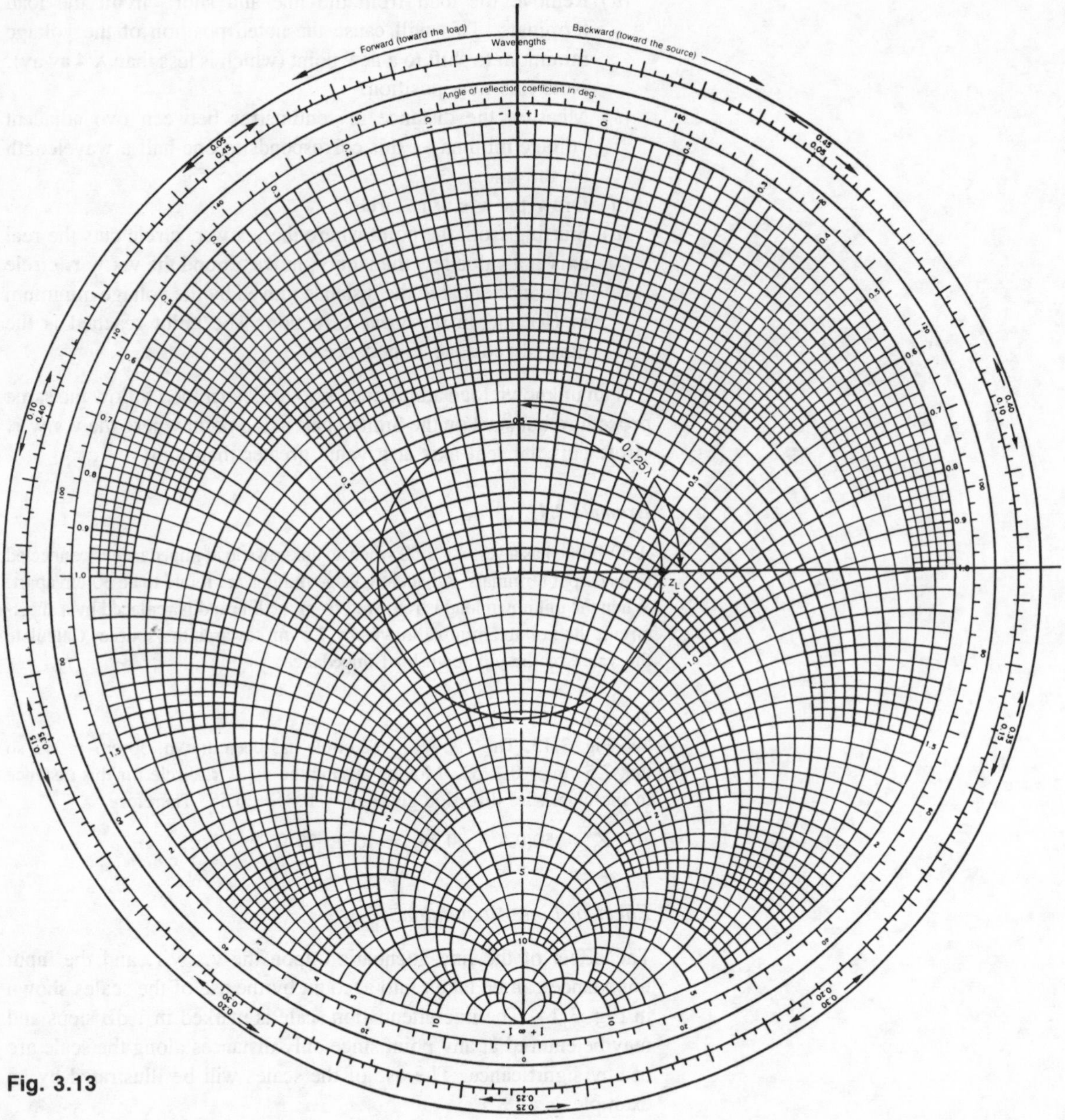

Fig. 3.13

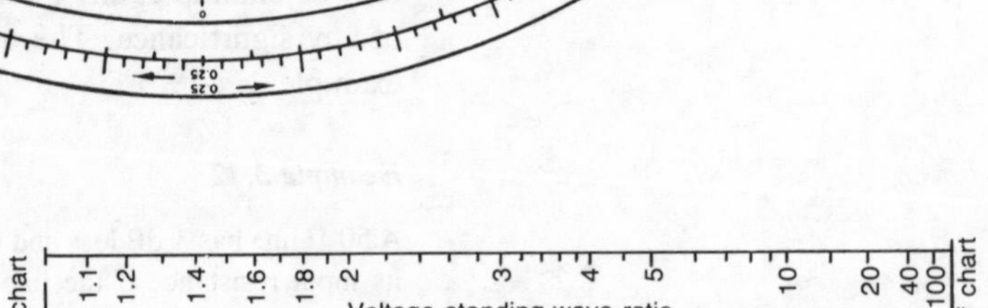

Fig. 3.14 Scales used to take into account any line attenuation.

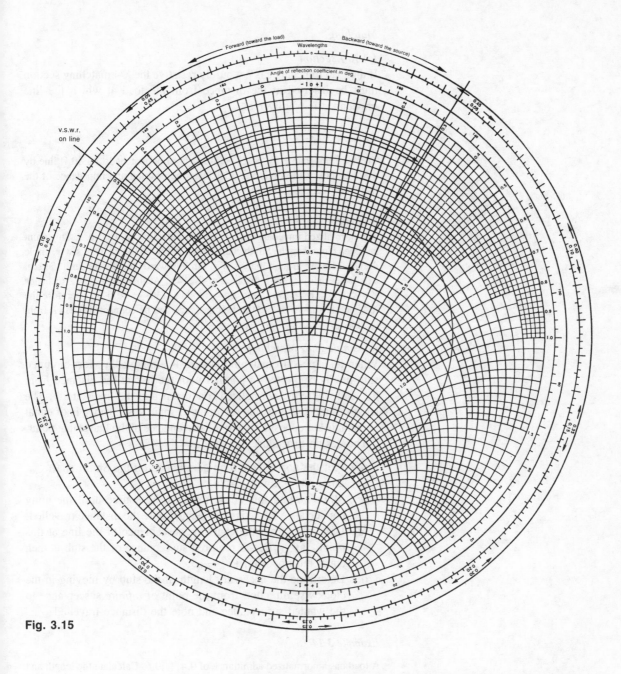

Fig. 3.15

This point is located on the chart and the loss-free v.s.w.r. circle drawn to give $S = 4$. Locate $S = 4$ on the v.s.w.r. scale of Fig. 3.14 and move across onto the line attenuation scale; this point is slightly less than 2.25 dB. Move along this scale towards the source a distance equal to 3 dB and then move back onto the v.s.w.r. scale. This gives $S = 1.89$ (approximately). The v.s.w.r. circle therefore becomes a spiral, as shown. The point reached is $z_{in} = 0.54 + j0.22$ so that $Z_{in} = 27 + j11\ \Omega$.   (*Ans.*)

Matching

(a) λ/4 section

If the load impedance is not purely resistive the λ/4 matching section must be inserted at the distance from the load at which the line impedance is wholly real.

Example 3.13

The load impedance $Z_L = 20 + j10 \ \Omega$ is to be matched to a 50 Ω line by a λ/4 length of loss-free line. Calculate the position and impedance of the matching section.

Solution

The normalized impedance $z_L = (20 + j10)/(50) = 0.4 + j0.2$. This point has been plotted on the Smith chart shown in Fig. 3.16 and the v.s.w.r. circle drawn. Moving around the v.s.w.r. circle towards the source until the real axis is reached covers a distance of 0.214λ. At this point $z = 2.6$; hence $Z = 50 \times 2.6 = 130 \ \Omega$ and $Z_0^1 = \sqrt{(50 \times 130)} = 80.6 \ \Omega$. Therefore the λ/4 section should be of 80.6 Ω impedance and be connected 0.214λ from the load.   *(Ans.)*

(b) Single-stub Matching

When a single-stub matching system is employed a short-circuited stub line is connected in parallel with the line at the distance from the load where the normalized conductance of the line is unity. The length of the stub must be such that its input susceptance is of equal magnitude but opposite sign to the susceptance of the line at that point.

The design procedure is as follows.

(*a*) Plot the normalized load admittance on the Smith chart.
(*b*) Draw the v.s.w.r. circle.
(*c*) Move around this circle towards the source until the unity conductance circle is reached and note the distance travelled.
(*d*) Determine from the chart the susceptance of the line at this point, say $+jB$; the required susceptance of the stub is then equal but opposite, i.e. $-jB$.
(*e*) Determine the necessary length of the stub by moving in the clockwise direction from the point of *infinite* susceptance to the wanted susceptance and note the distance travelled.

Example 3.14

A load has a normalized admittance of $0.4 + j0.6$. Calculate the length and position of a single stub used to match the load to the line.

Solution

The normalized load admittance y_L is plotted on the Smith chart shown in Fig. 3.17 and the v.s.w.r. circle drawn. The distance that must be travelled around the circle to reach the unity conductance circle is 0.078λ.   *(Ans.)*

At this point the normalized susceptance of the line is $+j1.35$. To find the required length of the stub move from the point $-j1.35$ on the edge of

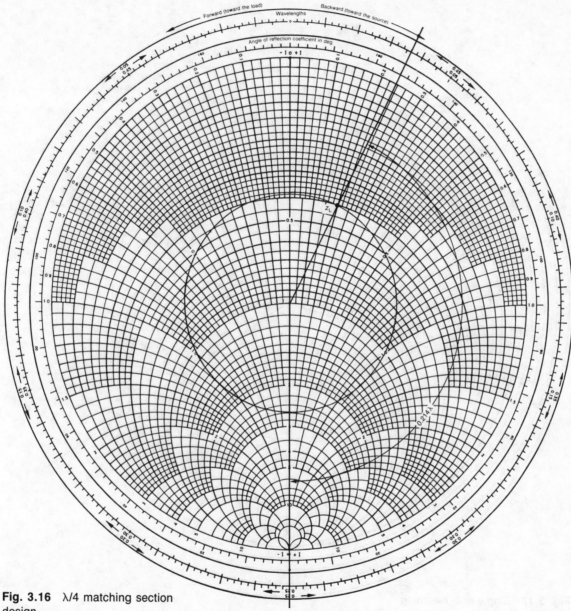

Fig. 3.16 λ/4 matching section design.

the chart to the point of infinite susceptance. This gives the length of the stub as 0.102λ.   (*Ans.*)

A single-stub matching system is satisfactory for single-frequency operation with a constant value of load impedance, but it is of limited usefulness otherwise.

(c) Double-stub Matching
If two short-circuited stubs are connected across a mismatched line

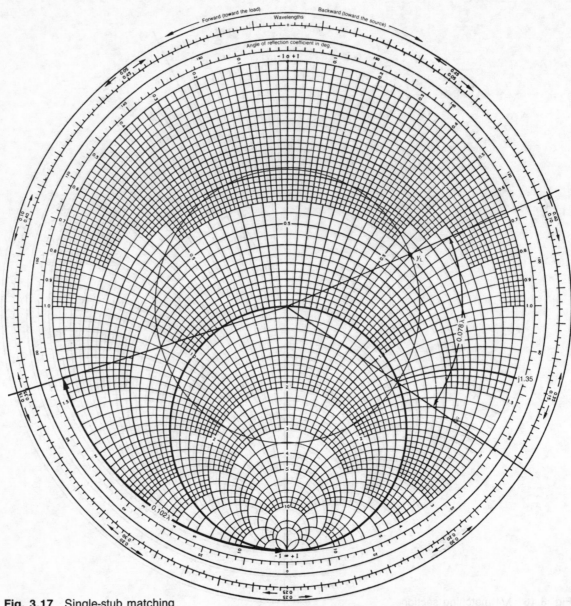

Fig. 3.17 Single-stub matching design.

at two different locations a much greater range of load impedances can be successfully matched together. The basic concept is illustrated by Fig. 3.18, the distance d_1 of the stub nearest the load may sometimes be zero. The stub nearest the source is only used to cancel out the susceptance of the line at that point, so that the normalized admittance of the line at this point is $y_1 = 1 + j0$ and hence $Y_1 = Y_0$. The stub nearest the load must have an input susceptance B_2 such that the total normalized admittance y_t ($y_t = Y_2 + jB_2/Y_0$) at the point of connection, when transformed by the length of line d_2, to become y_{in}, falls somewhere on the unity conductance circle.

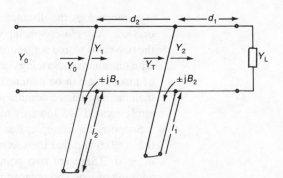

Fig. 3.18 Double-stub matching.

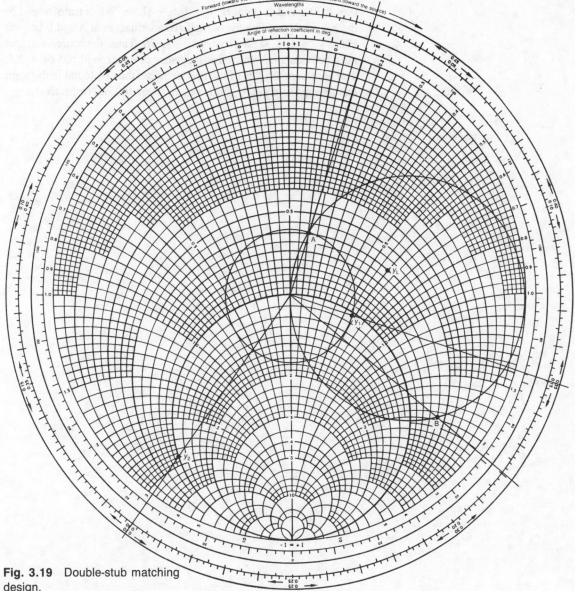

Fig. 3.19 Double-stub matching design.

In most cases the distance d_2 between the two stubs is either $\lambda/8$ or $3\lambda/8$. Any closer spacing will distort the field distribution so that the two stubs would act, more or less, as though they were in parallel with one another. Spacings between $\lambda/8$ and $3\lambda/8$ reduce the number of loads that can be matched. Suppose that d_2 is chosen to be $\lambda/8$; then the normalized admittance y_t must lie on the unity conductance circle rotated $\lambda/8$ towards the load, see Fig. 3.19.

Suppose, for example, that the normalized load admittance is $y_L = 0.6 + j0.6$ and that the second stub is connected at the load so that $d_1 = 0$. There are two points, marked as A and B, at which the addition of load susceptance will make the resultant load admittance lie on the $\lambda/8$ rotated unity conductance circle. At point A, $y_t = 0.6 + j0.1$ and at point B, $y_t = 0.6 + j1.96$. When transformed by the $\lambda/8$ section of line the normalized admittances at A and B become either $y_1 = 1 + j0.58$ or $y_2 = 1 - j2.76$. Thus, the source-end stub must provide a normalized susceptance of either $-j0.525$ or $+j2.6$. The necessary lengths of the two stubs can then be found in the same manner as for a single stub, i.e. 0.4λ or 0.05λ, respectively.

4 Waveguides

When a signal is propagated along a transmission line electromagnetic energy is guided in the space between the two conductors. Some of the energy is always lost because of the inevitable power dissipation in the resistance of the conductors. Even with a coaxial cable these losses become excessively high at frequencies above about 3 GHz. At these frequencies it becomes necessary to employ *rectangular waveguides* as the transmission medium. The rectangular waveguide, shown in Fig. 4.1, consists of a hollow, rectangular metal tube that has two horizontal walls of internal dimensions a and two vertical walls of internal dimensions b. Usually, the waveguide is made of copper, although sometimes other materials, such as aluminium or brass, are used. The dimensions a and b of the waveguide must be comparable with the wavelength of the signal and it is only at frequencies in excess of about 1 GHz that these dimensions become small enough for the use of a waveguide to be both practical and economic.

The use of a waveguide is made possible by the skin effect which limits the flow of current to the surface of a metal when the frequency is high enough. Energy can therefore be completely confined to the interior of the guide. Electromagnetic energy directed into the guide cannot radiate sideways and so it is propagated down the waveguide with little attenuation. An electromagnetic wave has both an electric and a magnetic component that are always in phase quadrature, and also mutually at right angles to the direction of propagation. This is shown by Fig. 4.2. When an electromagnetic wave is incident upon a perfect conductor no energy can be absorbed by the conductor and so the wave must be completely reflected. At the surface of the conductor the following boundary conditions must be satisfied.

(a) The electric field at the surface of the conductor must be perpendicular to the surface: there can be no tangential component. This must be so or else the field would be short-circuited by the (assumed) perfect conductor.
(b) Any magnetic field at the surface of the conductor must be parallel to the surface: there cannot be any perpendicular component.

The *polarization* of an electromagnetic wave is the plane in which the electric field lies; this is usually either the horizontal or the vertical plane but less often the polarization may be circular.

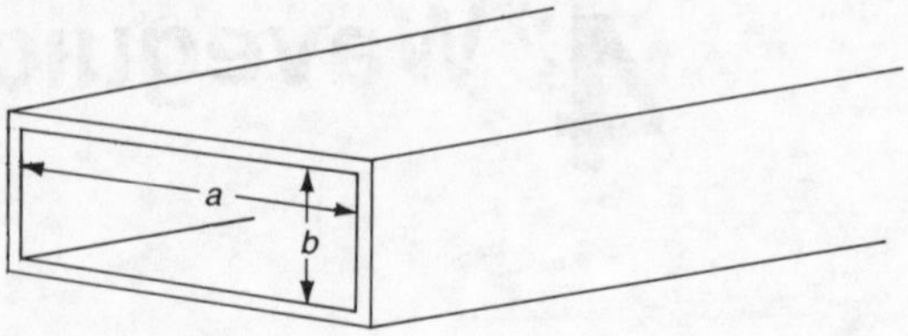

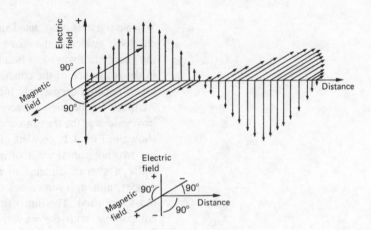

Fig. 4.1 Rectangular waveguide.

Fig. 4.2 Electromagnetic wave.

Propagation in the Rectangular Waveguide

Figure 4.3 shows one point in the positive peak wavefront of an electromagnetic wave entering a rectangular waveguide. The wave is incident upon one of the two vertical walls. The polarization of the wave is assumed to be such that the electric field is perpendicular to the plane of the paper and the magnetic field lies in the plane of the paper. When the electric field arrives at a wall it will be completely reflected with 180° phase reversal. This is indicated in the figure by the + and − signs. The total field at the surface of the wall is the phasor sum of the incident and reflected fields and it is zero. The angle of reflection θ is always equal to the angle of incidence θ. The reflected wave travels across the waveguide to the other vertical wall and here it is again totally reflected with another reversal in its polarity. This reflected wave travels across the waveguide to the other vertical wall, is again reflected, and so on. The electromagnetic wave will therefore propagate down the waveguide by means of a series of reflections from each of the vertical guide walls.

The wave travels from one vertical wall to the other with a velocity equal to the velocity of light c, i.e. 3×10^8 m/s. This velocity can be resolved into two components, as shown by Fig. 4.4. The *group velocity* V_G is the component of velocity parallel to the walls of the waveguide and it is the velocity with which the electromagnetic energy propagates down the waveguide. From Fig. 4.4, $V_G = c \cos \theta$ and this means that the group velocity is always less than the velocity of

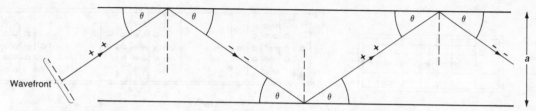

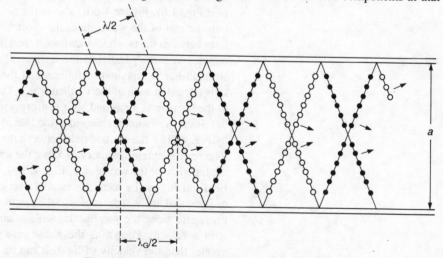

Fig. 4.3 Single point on a wavefront propagating down a waveguide.

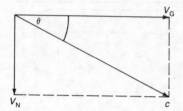

Fig. 4.4 Velocities in a waveguide.

light. The component of velocity V_N normal to the guide walls is equal to $c \sin \theta$. Figure 4.3 shows one particular point on the wavefront of a propagating electromagnetic wave. If the complete wavefront is considered the situation is somewhat more complex and it is illustrated by Fig. 4.5. As each successive peak of the wavefront travels across the waveguide, one end of the wavefront will arrive at a wall some time before the other end. As each part of the wavefront reaches the wall it will be immediately reflected, with a change in its polarity, whilst the remainder of the wavefront continues to move in its original direction. Eventually, the entire wavefront will have reached the wall and have been reflected to travel in the new direction. Figure 4.5 shows only the complete wavefronts; clearly, alternate positive and negative wavefronts propagate in both directions across the waveguide.

The distance between two successive wavefronts propagating in the same direction is equal to one-half of the free-space wavelength λ ($\lambda = c/f$). The distance between the intersection of two positive wavefronts and the intersection of two negative wavefronts is equal to one-half of the waveguide wavelength λ_G. These two different wavelengths are indicated on Fig. 4.5.

At every point within the waveguide the electric and the magnetic field strengths are the algebraic sum of the two components at that

Fig. 4.5 Wavefronts propagating down a waveguide (● = positive, ○ = negative).

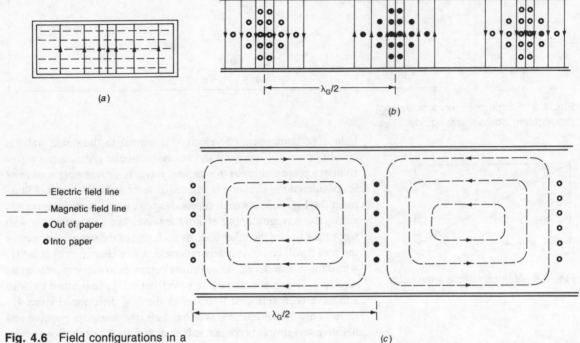

Electric field line
— — — Magnetic field line
● Out of paper
○ Into paper

Fig. 4.6 Field configurations in a rectangular waveguide, (a) end view, (b) side view, and (c) top view.

point. Whenever the peak of one of the waves intersects with a peak of another wave of the same polarity the resultant field will have its maximum value. Wherever two peaks of opposite polarity coincide they cancel out and the resultant field is zero; for the electric field this only occurs at the vertical walls of the waveguide. In this way the resultant electric and magnetic field configurations can be deduced (see Fig. 4.6). Figure 4.6(a) shows the field configurations looking into the end of the waveguide; the electric field consists of straight lines between the two horizontal walls, and the magnetic field consists of lines of force parallel to the horizontal walls. The electric field strength has its maximum amplitude at the centre of the guide and falls to zero at each of the vertical walls. The electric field is normal to the horizontal walls and so it conforms to the boundary condition (a). If the viewpoint is changed to the side of the waveguide, as shown by Fig. 4.6(b), the electric field between the horizontal walls changes its direction at regular intervals along the length of the guide. Lastly, the top view of the waveguide, Fig. 4.6(c), shows only the magnetic field and this can be seen to consist of a series of closed loops of lines of force that are at right angles to the electric field. The field patterns propagate along the waveguide, without any change in their shape, with a velocity known as the *phase velocity* V_P, which is always greater than the velocity of light. It can be seen from Fig. 4.6 that, while the magnetic field has a component which lies in the direction of propagation, the electric field does not. For this reason this *mode* of propagation is known as a *transverse electric mode* and it is labelled as TE_{10}. The first subscript 1 denotes that there is only one half-cycle in the field configurations in the *a* dimension of the waveguide. The second subscript 0 denotes that there are no half-cycle variations in the field strengths in the *b* dimension of the guide.

Other modes of propagation are also possible but the TE_{10} mode is both the easiest to set up and of the lowest frequency. The TE_{10} mode is by far the most commonly employed and it is often called the *dominant mode*. The higher-order modes may be generated whenever energy is delivered to, or taken from, the waveguide and also at any point where some field distortion occurs. Usually, the guide dimension *b* is chosen to ensure that all modes other than the dominant mode are suppressed. Usually, *b* is about one-half of *a*.

If the frequency of the signal is varied, the angle of incidence, and hence of reflection, will also vary to ensure that the dimension *a* is always equal to an integer number of half-wavelengths. If the frequency is increased the angle θ becomes larger and eventually the point is reached where only one half-wavelength is no longer possible and two become established. This would be the higher mode TE_{20}. On the other hand, if the frequency is decreased the angle θ becomes smaller and eventually the point is reached, when $\theta = 0°$, where the wave merely moves perpendicularly between the two walls of the waveguide. There is then *no* propagation of energy down the wave-guide; this means that a rectangular waveguide acts like a low-pass filter whose cut-off frequency is determined by the dimension *a*. The free-space wavelength λ_C at which this happens is known as the *cut-off wavelength*. The *cut-off frequency* $f_C = c/\lambda_C$ is the *lowest* frequency that the waveguide is able to transmit.

Refer to Fig. 4.7 which is an expanded version of the pattern between a positive peak and a negative peak. From the triangle ABC

$$\cos \theta = \frac{\lambda_G/2}{\sqrt{\left[a^2 + \left(\dfrac{\lambda_G}{2} \right)^2 \right]}},$$

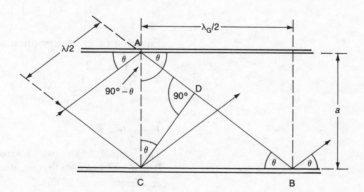

Fig. 4.7 Calculation of group wavelength.

from the triangle ADC

$\cos \theta = \lambda/2a,$

and from the triangle BCD

$\sin \theta = \lambda/\lambda_G.$

Therefore,

$$\frac{\lambda_G/2}{\sqrt{\left[a^2 + \left(\dfrac{\lambda_G}{2}\right)^2\right]}} = \frac{\lambda}{2a}, \quad \frac{\lambda_G^2 a^2}{4} = \frac{\lambda^2}{4}\left(a^2 + \frac{\lambda_G^2}{4}\right),$$

$$\lambda_G^2\left(a^2 - \frac{\lambda^2}{4}\right) = \lambda^2 a^2, \quad \lambda_G^2 = \frac{4\lambda^2 a^2}{4a^2 - \lambda^2} = \frac{\lambda^2}{1 - \dfrac{\lambda^2}{4a^2}},$$

or $\quad \lambda_G = \dfrac{\lambda}{\sqrt{\left[1 - \left(\dfrac{\lambda}{2a}\right)^2\right]}}.$ $\hspace{2cm}$ (4.1)

If $\lambda/2a < 1$, λ_G is real and the wave is able to propagate along the waveguide. If, however, $\lambda/2a > 1$, λ_G will be an imaginary quantity and this means that no propagation of energy takes place. The cut-off wavelength λ_C is the wavelength at which λ_G changes from a real to an imaginary quantity, i.e. $\lambda_C = 2a$. The cut-off wavelength of the next higher mode is equal to a. If the operation of the waveguide is to be restricted to the dominant mode only, the waveguide must be operated at some frequency between the cut-off frequencies of the two modes.

The lowest frequency for which the dominant mode is able to propagate down a waveguide is the frequency at which the wider dimension a is equal to one-half a wavelength. Therefore

$$\lambda_G = \frac{\lambda}{\sqrt{\left[1 - \left(\dfrac{\lambda}{\lambda_C}\right)^2\right]}}. \hspace{2cm} (4.2)$$

Equation (4.2) is very often written in the form

$$\frac{1}{\lambda_G^2} = \frac{1}{\lambda^2} - \frac{1}{\lambda_C^2}. \hspace{2cm} (4.3)$$

Example 4.1

The internal dimensions of a rectangular waveguide are 0.7112 cm by 0.3556 cm. Calculate (*a*) the cut-off frequency and (*b*) the group wavelength at a frequency of 25 GHz.

Solution

$\lambda_C = 2a = 2 \times 0.7112 = 1.4224$ cm.

(a) $f_C = \dfrac{c}{\lambda_C} = \dfrac{3 \times 10^8}{14.224 \times 10^{-3}} = 21.09$ GHz.   (*Ans.*)

(b) From equation (4.3)

$$\dfrac{1}{\lambda_G^2} = \left(\dfrac{25 \times 10^9}{3 \times 10^8}\right)^2 - \dfrac{1}{(14.224 \times 10^{-3})^2} \simeq 2000.$$

Hence $\lambda_G^2 = 5 \times 10^{-4}$ and $\lambda_G = 2.237$ cm.   (*Ans.*)

It is normally recommended that a rectangular waveguide should be used with signals having a free-space wavelength within ±20% of $4a/3$. The nominal operating frequency should be $1.5f_C$ with a recommended bandwidth of $1.25f_C$ to $1.9f_C$. This will ensure that only the dominant mode is transmitted. The next higher modes have cut-off wavelengths of a (TE_{20}) and $2b$ (TE_{01}). Hence, if $\lambda/2 < a < \lambda$ and $2b < \lambda$ these nearest modes will both be suppressed. Usually, b is approximately one-half of a so that the TE_{01} mode is suppressed and the attenuation of the waveguide is as low as possible.

The electromagnetic field components of the dominant mode at any point in a rectangular waveguide are given by the following equations.

$$E_x = E_z = H_y = 0$$

$$E_y = A \sin\left[\dfrac{\pi x}{a}\right] \sin(\omega t - \beta z)$$

$$H_x = -A\left[\dfrac{\beta}{\omega \mu_0}\right] \sin\left[\dfrac{\pi x}{a}\right] \sin(\omega t - \beta z) \qquad (4.4)$$

$$H_z = A\left[\dfrac{\pi}{a\omega \mu_0}\right] \cos\left[\dfrac{\pi x}{a}\right] \cos(\omega t - \beta z),$$

where x is distance in the a dimension, y is distance in the b dimension, z is distance along the length of the waveguide, β is the phase-change coefficient

$$\beta = \sqrt{\left[\left(\dfrac{\omega}{c}\right)^2 - \left(\dfrac{\pi}{a}\right)^2\right]} = \dfrac{2\pi}{\lambda_G},$$

and μ_0 is the permeability.

Velocity of Propagation

The *phase velocity* V_P is the rate at which any particular point in a field pattern moves along the waveguide. The *group velocity* V_G is the rate at which energy is propagated along the waveguide. From Fig. 4.8 the point on a wavefront marked as E will propagate along the waveguide to reach the point F first, and then G. To keep the

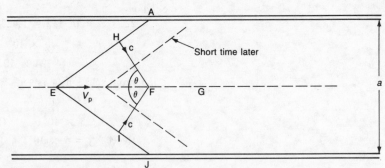

Fig. 4.8 Calculation of group and phase velocity.

phase of the wave constant the point E must arrive at F at the same time as any other point on the wavefront. The two wavefronts AE and EJ move in the directions perpendicular to the wavefronts at the velocity of light c. Their point of intersection E, a particular phase in the field pattern, moves along parallel to the walls at phase velocity V_P. For the two wavefronts *and* the point E to arrive at the point F simultaneously

$$\frac{c}{V_P} = \frac{HF}{EF} = \cos \theta.$$

Therefore

$$V_P = \frac{c}{\cos \theta}. \tag{4.5}$$

Since $\cos \theta < 1$ the phase velocity is always greater than, or equal to, the velocity of light. Further

$$V_P = f\lambda_G = \frac{c\lambda_G}{\lambda}. \tag{4.6}$$

The group velocity is the velocity with which energy is transmitted and it is equal to

$$V_G = c \cos \theta = \frac{c\lambda}{\lambda_G}. \tag{4.7}$$

Also

$$V_P V_G = c^2. \tag{4.8}$$

Example 4.2

A rectangular waveguide has internal dimensions 3.485 cm and 1.580 cm. Calculate the phase and group velocities at a frequency of 7 GHz.

Solution
$\lambda_C = 2a = 2 \times 3.485 = 6.97$ cm. Hence

$$\frac{1}{\lambda_G^2} = \left(\frac{7 \times 10^9}{3 \times 10^8}\right)^2 - \frac{1}{(69.7 \times 10^{-3})^2}$$

or $\lambda_G = 5.434$ cm.

From equation (4.6)

$$V_P = 7 \times 10^9 \times 5.434 \times 10^{-2} = 3.804 \times 10^8 \text{ m/s.} \quad (Ans.)$$

From equation (4.8)

$$V_G = \frac{(3 \times 10^8)^2}{3.804 \times 10^8} = 2.3659 \times 10^8 \text{ m/s.} \quad (Ans.)$$

Impedance of a Rectangular Waveguide

The ratio of the electric and the magnetic field strengths in a rectangular waveguide is known as the *wave impedance* Z_w and it is constant at all points in the guide. Thus,

$$Z_w = \left|\frac{E_y}{H_x}\right| = \frac{\omega\mu_0}{\beta} = \frac{2\pi c\mu_0/\lambda}{2\pi/\lambda_G} = c\mu_0 \cdot \frac{\lambda_G}{\lambda}$$

$$= \frac{\mu_0}{\sqrt{\mu_0\epsilon_0}} \frac{\lambda_G}{\lambda} = \sqrt{\left(\frac{\mu_0}{\epsilon_0}\right)} \frac{\lambda_G}{\lambda}$$

$$= 120\pi\lambda_G/\lambda. \tag{4.9}$$

Since $\lambda_G > \lambda$ the wave impedance is always greater than 377 Ω. If the wave impedance is matched to the waveguide's load then maximum power transfer from waveguide to the load will be obtained. If, however, the load is mismatched, reflections and hence standing waves, will be set up in the guide. The effect is exactly the same as occurs in a transmission line, and problems can be solved similarly.

Attenuation in a Rectangular Waveguide

As a signal is propagated down a waveguide the magnetic field induces e.m.f.s into the walls of the guide and these cause currents to flow. The direction of the current flow is always at right angles to the direction of the magnetic field adjacent to the wall. In the side walls (dimension b) the current flows vertically but in the horizontal walls the current distribution is as shown by Fig. 4.9. Because the walls must possess some resistance this flow of current results in power dissipation and, since this power can only be supplied by the propagating wave, attenuation. Some more losses are also introduced because the surfaces of the inner walls are not perfectly smooth. For all sizes of rectangular waveguide the variation of attenuation with frequency is of the form shown by Fig. 4.10. At the low-frequency end of the range the attenuation is high because of the cut-off effect mentioned earlier. There is then a relatively wide bandwidth over which the attenuation has a fairly flat minimum value rising only slowly with frequency, and then at higher frequencies the attenuation rises rapidly. Figures for each size of waveguide are given in Table 4.1.

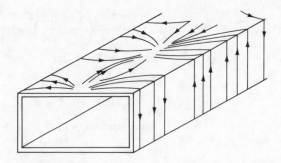

Fig. 4.9 Currents in waveguide walls.

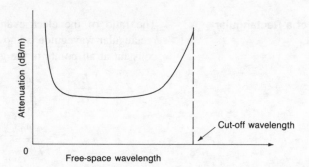

Fig. 4.10 Attenuation–frequency characteristics of a rectangular waveguide.

Table 4.1

| Waveguide dimensions (cm) | | | | Frequency limit (GHz) | | Cut-off frequency (GHz) | Power rating at 1.5f_c (MW) | Attenuation (dB/m) | WG | WR |
| Internal | | External | | | | | | | | |
a	b	a	b	Lower	Upper					
16.51	8.255	16.92	8.661	1.14	1.73	0.908	13.47	5×10^{-3}	6	650
10.92	5.461	11.33	5.867	1.72	2.61	1.373	5.90	9.4×10^{-3}	8	430
7.214	3.404	7.62	3.81	2.60	3.95	2.080	2.43	0.018	10	284
4.755	2.215	5.08	2.54	3.94	5.99	3.155	1.04	0.034	12	187
3.485	1.580	3.81	1.905	5.38	8.18	4.285	0.544	0.056	14	137
2.850	1.262	3.175	1.588	6.58	10.0	5.260	0.355	0.077	15	112
2.286	1.016	2.54	1.270	8.2	12.5	6.560	0.229	0.106	16	90
1.580	0.7899	1.783	0.9931	11.9	18.0	9.49	0.123	0.171	18	62
1.067	0.4318	1.27	0.635	17.6	26.7	14.08	0.048	0.357	20	42
0.7112	0.3556	0.9144	0.5588	26.4	40.1	21.1	0.025	0.5767	22	28
0.569	0.2845	0.7722	0.4877	33.0	50.1	26.35	0.016	0.787	23	22
0.4775	0.2388	0.6807	0.4420	39.3	59.7	31.4	0.010	1.026	24	19
0.3759	0.1880	0.5791	0.3921	49.9	75.8	39.9	0.007	1.466	25	15
0.3099	0.1549	0.5131	0.3581	60.5	92.0	48.4	0.005	1.957	26	12

Power Handling Capability

The maximum power that can be transmitted by a rectangular wave-guide is limited by the need to avoid voltage breakdown of the air within the guide. This means that it is necessary to ensure that the maximum electric field strength is always less than the breakdown value; for dry air at atmospheric pressure this is 3×10^6 V/m. The maximum power that a waveguide can transmit using the dominant mode is given by equation (4.10), i.e.

$$P_{max} = \frac{E^2_{max}ab}{480\pi} \sqrt{\left[1 - \left(\frac{\lambda}{2a}\right)^2\right]}. \qquad (4.10)$$

Sizes of Waveguides

A range of standard sizes of rectangular waveguides has been developed and these are given by Table 4.1. For each size there are recommended upper and lower frequency limits, a power rating and the attenuation at 1.5 times the cut-off frequency. The microwave spectrum is often said to be divided into a number of bands each of which is given a label. Unfortunately the labelling is not universally agreed but Fig. 4.11 shows one version that is employed in the USA.

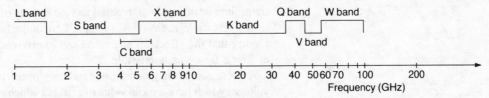

Fig. 4.11 Microwave bands. L 0.39 − 1.55, S 1.55 − 5.2, X 5.2 − 10.9, K 10.9 − 36, Q 36 − 46, V 46 − 56, and W 56 − 100 GHz.

Two other microwave labelling schemes are also in use: (a) United Kingdom IEE; C 0.5 − 1, D 1 − 2, E 2 − 3, F 3 − 4, G 4 − 6, H 6 − 8, I 8 − 10, J 10 − 20, K 20 − 40, L 40 − 60, and M 60 − 100 GHz. (b) NATO; L 1 − 2, S 2 − 4, C 4 − 8, X 7 − 12, J 12 − 18, K 18 − 26, Q 26 − 40, V 40 − 60, and O 60 − 90 GHz.

5 Noise in Radio Systems

Noise and interference, which may be regarded as any unwanted signal that appears at the output of a radio system, set a lower limit to the usable signal level at that output. For the output signal to be useful the signal power must be larger than the noise power by an amount specified by the required minimum signal-to-noise ratio. This, in turn, varies considerably depending upon the nature of the signal; two examples of minimum signal-to-noise ratios are: (*a*) 30 dB for a telephone circuit, and (*b*) 60 dB for a music circuit. The required minimum signal-to-noise ratio determines the power that a radio transmitter must supply to its aerial and the spacing of the relay stations in a microwave radio-relay system. Economic factors, therefore, require that the effects of any noise and interference are reduced to as low a figure as is possible.

A noise source may consist of randomly occurring, non-periodic voltages which have a mean value of zero but which may contain some relatively large voltage peaks. Such noise sources have a uniform power density over a given bandwidth and are said to be *white*; it is possible to calculate the r.m.s. value of a white-noise source. Other noise sources are impulsive in their nature: they usually originate from various man-made sources such as electric motors, neon signs, and so on. Interfering signals may be picked up by a receive aerial and if inadequately filtered may appear at the output of the receiver; the main culprits here are the adjacent-channel and image-channel signals (p. 191). Other interfering signals may be generated within the radio receiver itself as a direct result of *intermodulation* (p. 192).

Sources of Noise in Radio Systems

The various sources of noise that may arise and degrade the performance of a radio system may conveniently be divided into two groups. These are: (*a*) noise generated within the radio receiver itself, and (*b*) noise that is picked up by the receive aerial; this, in turn, may originate from either natural or man-made sources.

Noise Generated within a Radio Receiver

Thermal Noise

When the free electrons in a conductor receive heat energy they will be caused to move randomly about in that conductor. The movement of an electron constitutes a current which gives rise to a random noise voltage that appears across the conductor. The r.m.s. value of this voltage is

$$V_n = \sqrt{(4kTBR)} \text{ V},\qquad(5.1)$$

where: k is the Boltzmann constant ($= 1.38 \times 10^{-23}$ J/K); T is the absolute temperature in K (K = °C + 273); B is the *noise* bandwidth in Hz; and R is the resistive part of the impedance in which the noise is generated.

The standard reference temperature is usually taken as being 290 K and is given the symbol T_0. This temperature should always be assumed unless some other value is given.

A resistance R ohms can be represented by a noise voltage generator V_n in series with a noiseless resistor of R ohms, as shown by Fig. 5.1. The *noise bandwidth* of a circuit, or device, is the width along the frequency axis of a rectangular response curve whose area and height are the same as the actual frequency response characteristic of the circuit (see Fig. 5.2). The noise bandwidth of a circuit can be determined graphically or by the use of the expression

$$B_n = \int_0^\infty \frac{|A_v(f)|^2 \, \mathrm{d}f}{|A_{v(\text{max})}|^2},\qquad(5.2)$$

where $A_v(f)$ is the voltage gain of the circuit at any frequency f, and $A_{v(\text{max})}$ is the maximum voltage gain.

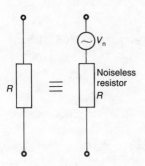

Fig. 5.1 Representation of the thermal noise generated in a resistor.

Example 5.1

Derive an expression for the noise bandwidth of the circuit shown in Fig. 5.3. Calculate the noise bandwidth if the output voltage falls by 3 dB at a frequency of 12 kHz.

Solution

$$\frac{V_{\text{out}}}{V_{\text{in}}} = \frac{1/\mathrm{j}\omega C}{R + 1/\mathrm{j}\omega C} = \frac{1}{1 + \mathrm{j}\omega CR}$$

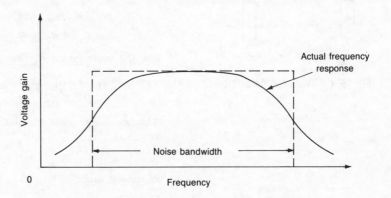

Fig. 5.2 Noise bandwidth.

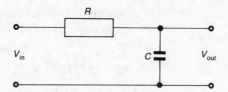

Fig. 5.3

$$\left|\frac{V_{out}}{V_{in}}\right| = \frac{1}{\sqrt{(1 + \omega^2 C^2 R^2)}}$$

and this falls by 3 dB at a frequency $\omega_0/2\pi$, where $\omega_0 = 1/CR$. Hence

$$A_v(f) = \frac{1}{\sqrt{(1 + f^2/f_0^2)}}.$$

From equation (5.2)

$$B_n = \int_0^\infty \frac{df}{1 + f^2/f_0^2} = f_0^2 \int_0^\infty \frac{df}{f_0^2 + f^2} \; \dagger$$

$$= f_0^2 \left[\frac{1}{f_0} \tan^{-1}\left(\frac{f}{f_0}\right)\right]_0^\infty = \frac{\pi f_0}{2}. \quad (Ans.)$$

If $f_0 = 12$ kHz, $B_n = 15\ 708$ Hz.   *(Ans.)*

In similar manner it can be shown that for (*a*) a single-tuned circuit $B_n = 1.57B_{3\ dB}$, and (*b*) two cascaded single-tuned circuits $B_n = 1.22B_{3\ dB}$. If the noise bandwidth of a circuit is not known the 3 dB bandwidth will have to be used; this will introduce some error of course but this error will be small if the gain–frequency characteristic of the circuit has a rapid roll-off.

Available Noise Power
If a resistor R_1 at temperature T_1 is connected to another resistor R_2 at temperature T_2 as shown by Fig. 5.4, noise power will be transferred from each resistor to the other resistor.

The noise power P_1 delivered by resistor R_1 to resistor R_2 is

$$P_1 = \left(\frac{V_{n1}R_2}{R_1 + R_2}\right)^2 \frac{1}{R_2}$$

and the noise power P_2 delivered from R_2 to R_1 is

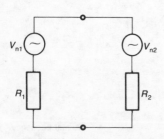

Fig. 5.4 Available noise power.

$$P_2 = \left(\frac{V_{n2}R_1}{R_1 + R_2}\right)^2 \frac{1}{R_1}.$$

The net noise power transferred from source to load is

†A standard integral is $\int \frac{dx}{x^2 + y^2} = \frac{1}{y}\tan^{-1}\left(\frac{x}{y}\right)$.

$$P_n = P_1 - P_2 = \frac{4kB(R_1 R_2 T_1 - R_1 R_2 T_2)}{(R_1 + R_2)^2}$$

$$\text{or} \quad P_n = \frac{4kB \, R_1 R_2 (T_1 - T_2)}{(R_1 + R_2)^2}. \tag{5.3}$$

The maximum, or *available*, noise power P_A occurs when the two resistors are of equal value and $T_2 = 0$ (i.e. resistor R_2 is noise free). Then

$$P_A = \left(\frac{V_n/2}{R}\right)^2 = \frac{4kTBR}{4R} = kTB \text{ W}. \tag{5.4}$$

If P_A is divided by the bandwidth B the available noise power in a 1 Hz bandwidth is obtained; this is known as the *power density spectrum* (p.d.s.) and it is constant up to about 300 GHz.

Very often it is convenient to quote the available noise power in decibels. Thus

$$P_A = -174 \text{ dBm} + 10 \log_{10} B_n. \tag{5.5}$$

Available Power Gain

There are a number of ways in which the power gain, or loss, of a network can be defined, but for noise calculations the *available power gain G* should be used. This is the ratio (output power)/(input power) when *both* the input and the output impedances of the network are matched to the source and the load, respectively.

$$\text{Available gain } G = \frac{\text{available output power}}{\text{available input power}}. \tag{5.6}$$

The expression for the noise bandwidth (5.2) can be written in terms of the available power gain

$$B_n = \int_0^\infty \frac{|G(f)| \, df}{G_{max}}, \tag{5.7}$$

where $G(f)$ is the available power gain at any frequency f, and G_{max} is the mid-band available power gain.

Noise Produced by Resistances in Series

If two, or more, resistances are connected in series the total mean-square noise voltage is the sum of the mean-square noise voltages generated by each resistance. For two resistances R_1 and R_2, the total mean-square noise voltage V_n^2 is $V_n^2 = V_{n1}^2 + V_{n2}^2 = 4kB(R_1 T_1 + R_2 T_2)$. Usually, the resistances are at the same temperature, and then $V_n^2 = 4kBT(R_1 + R_2)$, which means that the total mean-square noise voltage is effectively generated in the total resistance of the circuit.

Noise Produced by Resistors in Parallel

If two resistors are connected in parallel and are at the same temperature the total mean-square noise voltage will be equal to $4kTBR_T$, where $R_T = R_1 R_2/(R_1 + R_2)$. If the two resistors are not at the same temperature the superposition theorem will have to be used to calculate the total mean-square noise voltage generated.

Example 5.2

An amplifier has an input resistance of 2000 Ω and it is at a temperature of 290 K. It is connected to a source of 1000 Ω resistance at a temperature of 100 K. Calculate the total noise voltage across the input terminals of the amplifier if the noise bandwidth is 1 MHz.

Solution

The thermal noise voltages V_{ns} and V_{ni} generated in the source and input impedances respectively are

$$V_{ns} = \sqrt{(4 \times 1.38 \times 10^{-23} \times 100 \times 10^6 \times 10^3)} = 2.35 \ \mu V,$$

and $V_{ni} = \sqrt{(4 \times 1.38 \times 10^{-23} \times 290 \times 2 \times 10^9)} = 5.66 \ \mu V$.

Referring to Fig. 5.5, the total noise voltage across the amplifier's input terminals is

$$V_{nt} = \sqrt{\left[\left(\frac{2.35 \times 2}{3}\right)^2 + \left(\frac{5.66 \times 1}{3}\right)^2\right]} = 2.45 \ \mu V. \quad (Ans.)$$

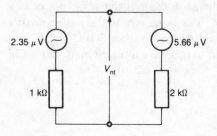

2.35 μV   5.66 μV

V_{nt}

1 kΩ 2 kΩ

Fig. 5.5

Semiconductor Noise

All semiconductor devices may be subject to the following sources of noise.

(a) *Thermal agitation noise* in the resistance of the semiconductor material occurs mainly in the base resistance of a bipolar transistor and in the channel resistance of an FET.

(b) *Shot noise* is caused by the random movement of electrons and holes across each p–n junction in the device. The r.m.s. value of the shot-noise current is given by $I_n = \sqrt{(2eIB)}$, where e is the electronic charge = 1.602×10^{-19} C, I is the d.c. current flowing across the junction and B is the noise bandwidth. Shot noise is white.

(c) *Flicker or 1/f noise* caused by fluctuations in the conductivity of the semiconductor material; it is inversely proportional to frequency.

Intermodulation Noise

A major cause of noise in radio systems carrying multi-channel analogue telephony signals is intermodulation (p. 192). Intermodulation occurs whenever a complex signal is applied to a non-linear device; it results in the production of components at frequencies equal to the sums and the differences of the frequencies, and at the harmonics of the frequencies, contained in the input signal. When the input signal contains components at several different frequencies the number of intermodulation products may be very large; the effect at the output of the circuit is then very similar to that of thermal agitation noise. The main difference between them is that whereas thermal agitation noise is constant, intermodulation noise is a function of the amplitude of the input signal.

Noise Picked up by the Receiving Aerial

The noise picked up by a receive aerial is generated by a number of sources of atmospheric, galactic, and man-made origin. At frequencies in the low-, medium-, and high-frequency bands the external noise is much larger than the noise generated within the radio receiver. At higher frequencies (in the v.h.f., u.h.f. and s.h.f. bands) externally generated noise falls to a relatively low level and the overall noise performance of a system is mainly determined by the receiver itself. Considerable attention is then given to reducing the noise factor of the receiver to as small a figure as is possible.

Natural Sources of Noise

A receiving aerial will pick up noise from the sky and from the earth itself as well as from various interfering signals. Sky noise has a magnitude that varies both with frequency and with the direction to which the aerial is pointed. Sky noise is normally expressed in terms of the *noise temperature* T_A of the aerial. This is the temperature at which the aerial must be assumed to be for thermal agitation in its radiation resistance (p. 115) to produce the same noise power as is actually supplied by the aerial. Thus $T_A = $ (noise power)$/kB$. If the aerial is used for terrestrial communications, so that its main beam has only a small upwards inclination then its noise temperature is effectively that of the earth, or of the lower atmosphere, and this is usually taken as being 300 K. If, on the other hand, the aerial points upwards to the sky its noise temperature may be that of space which is only a few kelvin.

At medium and high frequencies atmospheric noise or *static* is always present. Every time a flash of lightning occurs somewhere in the world impulse noise is generated. Since thunderstorms are always simultaneously occurring at different points around the earth and the noise generated can propagate for very long distances, atmospheric noise is always present. The combined effect of a great many noise impulses makes it sound at the output of a receiver very like thermal agitation noise. Atmospheric noise has its greatest magnitude (approximately 10 μV/m) at about 10 kHz and it is negligible at frequencies above about 20 MHz. The level of atmospheric noise varies considerably with the location of the aerial, with the time of day and year, and with the frequency.

The manner in which sky noise varies with frequency above

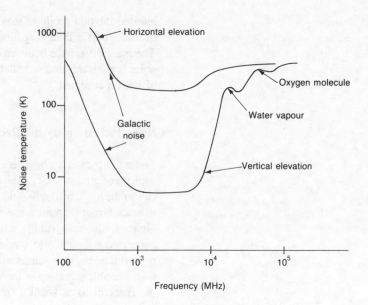

Fig. 5.6 Variation of sky noise with frequency.

100 MHz is shown, for both horizontal and vertical elevations, by Fig. 5.6. Galactic noise, which is produced by radiation from distant stars, has its maximum value at about 20 MHz and then it falls inversely proportional to (frequency)$^{2.4}$ to a minimum at about 1 GHz. At 50 MHz T_A is in the region of 5000 K. Galactic noise is negligible at frequencies higher than about 500 MHz and then the main source of noise is the earth itself. Noise due to radiation from the earth also decreases with increase in frequency and, typically, the aerial noise temperature T_A is about 300 K at 200 MHz. Another source of sky noise is the sun. The sun has a noise temperature of about 6000 K or more and occasionally produces bursts of noise which are several times greater than this. The effect of solar noise can always be minimized by pointing the aerial away from the sun. At higher frequencies the noise temperature rises with peaks at 23 GHz, due to water vapour molecules giving a peak in atmospheric absorption, and at about 60 GHz due to oxygen molecules.

Man-made Sources of Noise

There are a large number of possible sources of man-made noise which may be picked up by a receiving aerial. Whenever an electric current is switched on or off one, or more, voltage spikes will be generated because of the inevitable capacitance and inductance of all circuits. The voltage spike may be a single transient, or several transients may occur in rapid succession. Transient spikes occur whenever a circuit is switched on or off by a mechanical or an electronic switch; examples of such switching are vehicle ignition systems, electric light switches and the brushes in electric motors. These cause audible clicks that

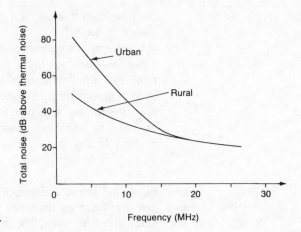

Fig. 5.7 Typical variation of total aerial noise in urban and rural areas.

may adversely affect reception and it is difficult to specify the actual circuit degradation. Continuous interference spectrums are generated by circuits such as thyristor speed controls and switched-mode power supplies.

The interference may either be electromagnetic (e.m.i.) or it may be radio-frequency (r.f.i.). The former is of lower frequency (30 kHz to 30 MHz) but it is usually of higher voltage and it may travel over the mains wiring for a considerable distance. Radio-frequency interference is of high frequency (30 to 300 MHz) and is both conducted over, and radiated from, the mains wiring which acts as a rather inefficient aerial. The total man-made noise picked up by a receive aerial cannot be separately identified from the atmospheric and galactic noise that is also present, but obviously this source of noise will be larger in an urban environment than in a rural area. Figure 5.7 shows how, typically, the total of atmospheric, galactic and man-made noise may vary in both urban and rural areas.

Man-made noise can often be suppressed at its source, although it may not always be economical to do so. Alternatively, it is often possible to position the receive aerial at a site which is remote from interference.

Signal-to-noise Ratio

As a signal is propagated over a radio link it will be subjected to various forms of noise and/or interference and it will also suffer attenuation. The radio receiver will have sufficient gain to compensate for the path attenuation and give the required output signal power, but the amplifier and mixer stages will add further noise to the signal. The usefulness of the output signal for its intended purpose is expressed by means of its *signal-to-noise ratio*. This is defined as

$$\text{signal-to-noise ratio} = \frac{\text{wanted signal power}}{\text{unwanted noise power}}. \qquad (5.8)$$

Because signal-to-noise ratio is the ratio of two powers it is commonly quoted in decibels.

The larger the output signal-to-noise ratio of a radio system the greater will be its ability to satisfactorily receive weak signals. The minimum permissible signal-to-noise ratio of a radio system is one of its most important parameters. Different minimum signal-to-noise ratio figures are specified for different kinds of signal and, for example, the minimum figure for a mobile radio-telephony system is about 15 dB.

Noise Factor

Figure 5.8 shows a circuit that has an available power gain G and that introduces an internally generated noise power at the output terminals of N_C watts. The input signal to the circuit is S_{in} with an associated noise power of $N_{in} = kT_0B$ watts. This means that the input signal-to-noise ratio is S_{in}/N_{in}.

The noise factor, or noise figure, F of the circuit is defined as

$$F = \frac{\text{total noise power at output}}{\text{amplified input noise power}}. \qquad (5.9)$$

It is assumed that both the source and the circuit are at the standard temperature of 290 K and that the circuit is linear. Therefore

$$F = \frac{GN_{in} + N_C}{GN_{in}} = 1 + \frac{N_C}{GN_{in}}. \qquad (5.10)$$

The noise factor is unaffected by the value of the load impedance because any mismatch at the output terminals which might exist will reduce both the noise and the signal powers equally, and so will not alter their ratio. If the circuit were noise free N_C would be zero and then the noise factor would be $F = 1$ or 0 dB. This is the theoretical minimum figure for the noise factor. In practice, typical figures are 4 dB for a v.h.f. amplifier and 12 dB for a radio receiver.

It is often convenient to refer the noise generated within the circuit to the input terminals. Then $N_e = N_C/G$ and

$$F = \frac{GN_{in} + GN_e}{GN_{in}} = 1 + \frac{N_e}{N_{in}}. \qquad (5.11)$$

This expression can be rearranged to give $N_e = (F - 1)N_{in}$, which expresses the internally generated noise as a function of both the noise factor and the actual input noise. The internally generated noise may

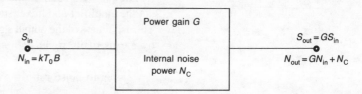

Fig. 5.8 Noise factor.

be regarded as being generated by thermal agitation in an equivalent input resistance (see *Electronics IV*, Ch. 8).

When a coherent input signal is applied to the circuit the expression for the noise factor can be written as

$$F = \frac{\left(\dfrac{1}{GN_{\text{in}}}\right)}{\left(\dfrac{1}{GN_{\text{in}} + N_{\text{C}}}\right)} = \frac{\left(\dfrac{1}{N_{\text{in}}}\right)}{\left(\dfrac{G}{GN_{\text{in}} + N_{\text{C}}}\right)}$$

$$= \frac{\left(\dfrac{S_{\text{in}}}{N_{\text{in}}}\right)}{\left(\dfrac{GS_{\text{in}}}{GN_{\text{in}} + N_{\text{C}}}\right)}$$

or $\quad F = \dfrac{\text{input signal-to-noise ratio}}{\text{output signal-to-noise ratio}}.$ (5.12)

This expression shows that the noise factor of a circuit indicates the extent to which an input signal-to-noise ratio is degraded by the noise generated internally within the circuit.

Output Noise from a Circuit

The available input noise power to a circuit is kT_0B watts, and putting this into equation (5.12) gives

$$F = \frac{S_{\text{in}}/kT_0B}{GS_{\text{in}}/N_0} = \frac{N_0}{GkT_0B},$$

where N_0 is the output noise power. Therefore

$$N_0 = FGkT_0B. \tag{5.13}$$

In decibels this is

$$N_0 = -174 \text{ dBm} + 10 \log_{10} B + 10 \log_{10} F + 10 \log_{10} G. \tag{5.14}$$

This is the total noise power at the output of the circuit and it is the sum of (*a*) the amplified input noise, and (*b*) the noise produced within the circuit. Since the former is equal to GkT_0B, the latter must be

$$N_e = (F - 1)GkT_0B. \tag{5.15}$$

In the case of a radio receiver which has different bandwidths at different points in the circuit the narrowest bandwidth should be used.

Variation of Noise Factor with Frequency

The gain of a circuit and the noise generated within it per unit bandwidth are often a function of frequency and this means that the noise factor of the circuit may also vary with frequency. Thus a distinction between the noise factor at a single frequency and the full-bandwidth noise factor may need to be made. If the bandwidth of the circuit is narrow enough for any variations in the gain and/or generated noise to be ignored, the *spot noise factor* is obtained. The spot noise factor can be measured at a number of points in the overall bandwidth of the circuit and then the *average noise factor* can be obtained.

Noise Factor of a Lossy Network

Suppose the network has an attenuation (power ratio) of X. The available input noise power is kT_0B and hence the available output noise power is $(kT_0B)/X + N_C$. This power is the same as the available noise power kT_0B from the output resistance of the network. Therefore

$$\frac{kT_0B}{X} + N_C = kT_0B$$

or $N_C = kT_0B\left(1 - \frac{1}{X}\right) = kT_0B\left(\frac{X-1}{X}\right).$

The same available output noise power would be obtained if a noiseless network having the same attenuation was to be supplied by a source that produces an available noise power of $kT_0B(X - 1)$. Hence

$$kT_0B(X - 1) = (F - 1)kT_0B$$

or $F = X.$ (5.16)

Thus the noise factor of a lossy network, such as an attenuator or a length of coaxial or waveguide feeder, at temperature T_0 is equal to its attenuation.

Overall Noise Factor of Circuits in Cascade

Figure 5.9 shows two circuits connected in cascade. The circuits have noise factors of F_1 and F_2, and available gains of G_1 and G_2, respectively, and it is assumed that their bandwidths are the same. If the overall noise factor of the combination is F_0 then the available output noise power N_0 will be equal to $F_0G_1G_2kT_0B$. For the first circuit, the output noise power is $N_{01} = F_1G_1kT_0B$ and this is the input noise to the second circuit. Hence

$$N_{02} = G_2F_1G_1kT_0B + (F_2 - 1)G_2kT_0B.$$

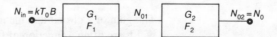

Fig. 5.9 Noise factors in cascade.

Since $N_0 = N_{02}$,

$$F_0 G_1 G_2 k T_0 B = G_1 G_2 F_1 k T_0 B + (F_2 - 1) G_2 k T_0 B$$

and $F_0 = F_1 + \dfrac{F_2 - 1}{G_1}.$ (5.17)

This result can be extended to any number of circuits connected in cascade. For n circuits

$$F_0 = F_1 + \frac{F_2 - 1}{G_1} + \frac{F_3 - 1}{G_1 G_2} + \ldots + \frac{F_n - 1}{G_1 G_2 \ldots G_{n-1}}.$$
(5.18)

If the first, and later, circuits in Fig. 5.9 are amplifiers the noise generated by the second, and following, amplifiers will be reduced because of its division by the product of the gains of the preceding amplifiers. If, however, the first circuit introduces a loss then the noise introduced by the second circuit will be increased. This means that to obtain a good, i.e. a low, overall noise factor, the first circuit *must* introduce the minimum possible loss and should, if possible, be an amplifier.

Example 5.3

A radio receiver has a noise factor of 6 dB. An amplifier with a power gain of 10 dB is connected between the aerial and the receiver. The overall noise factor is then 6 dB. Calculate the noise factor of the amplifier. Calculate the overall noise factor if a 6 dB attenuator were to be connected (*a*) between the aerial and the amplifier, and (*b*) between the amplifier and the receiver. Assume the noise picked up by the aerial to have an effective noise temperature of 290 K.

Solution
From equation (5.17)

$$3.98 = F_1 + \frac{3.98 - 1}{10}$$

or $F_1 = 3.682 = 5.66$ dB.   (*Ans.*)

$$(a)\ F_0 = 3.98 + \frac{3.682 - 1}{1/3.98} + \frac{3.98 - 1}{10 \times 1/3.98} = 15.84 = 12 \text{ dB}.$$

(*Ans.*)

$$(b)\ F_0 = 3.682 + \frac{3.98 - 1}{10} + \frac{3.98 - 1}{10 \times 1/3.98} = 5.17 = 7.13 \text{ dB}.$$

(*Ans.*)

Effective Noise Factor when $T_S \neq T_0$

The noise factor of a circuit is defined, measured and quoted with reference to a temperature of 290 K. If the source temperature is *not* at 290 K but is at some other temperature the degradation of the input signal-to-noise ratio will be different from that indicated by the noise factor. If the source temperature is less than 290 K the degradation will be greater than indicated, but if the source temperature is in excess of 290 K the signal-to-noise ratio reduction will be less than expected. To obtain the correct figure for the output signal-to-noise ratio the *effective noise factor* F_{eff} will have to be employed.

From equation (5.13), the noise factor of a circuit referred to the standard temperature of 290 K is $F = N_0/GkT_0B$, and hence the available output noise power due to the circuit alone is $(F - 1)GkT_0B$.

If, now, a source at temperature $T_S \neq T_0$ is applied to the input terminals of the network, the available noise output power will be $N_0 = GkT_SB + (F - 1)GkT_0B$. Now the effective noise factor $F_{eff} = N_0/GkT_SB$, or

$$F_{eff} = \frac{GkT_SB + (F - 1)GkT_0B}{GkT_SB} = 1 + \frac{T_0}{T_S}(F - 1).$$

(5.19)

Example 5.4

A radio receiver has a noise factor of 10 and it is connected by a feeder of 3 dB loss to an aerial. The noise delivered by the aerial is at an effective noise temperature of 145 K. Calculate the noise power at the output of the receiver if the receiver has a power gain of 50 dB and a bandwidth of 10 kHz.

Solution
The input noise power to the receiver system is

$$kT_AB = 1.38 \times 10^{-23} \times 145 \times 10^4 = 2 \times 10^{-17} \text{ W}.$$

The noise factor of the feeder is 3 dB, or 2, and so the overall noise factor is $2 + (10 - 1)/\frac{1}{2} = 20$. The effective noise factor at 145 K is

$$F_{eff} = 1 + \frac{(20 - 1)290}{145} = 39.$$

Therefore, the output noise power is

$$N_0 = FGkT_AB = 39 \times 10^5 \times 2 \times 10^{-17} = 78 \text{ pW}. \quad (Ans.)$$

Noise Temperature

The concept of noise temperature provides an alternative to noise factor for the specification of the noise performance of a radio system. The noise power N_C generated *within* a circuit and appearing at the output terminals may be considered to be the result of thermal agitation in the output resistance of the circuit. The output resistance must be regarded as being at a noise temperature $T_n = tT_0$, where t may be greater than, or smaller than, unity. Thus

$$N_C = kT_nB$$

or $\quad T_n = \dfrac{N_C}{kB}$. $\hspace{4cm}$ (5.20)

If the internally generated noise is referred to the input of the circuit it may be regarded as entering the input of the circuit as shown by Fig. 5.10. The internally generated noise is now equal to $kT_n B$ and so the total input noise is $kB(T_A + T_n)$, where T_A is the noise temperature of the aerial which also may, or may not, be equal to T_0.

The input signal-to-noise ratio of Fig. 5.10 is $S_{in}/kT_A B$ and the output signal-to-noise ratio is

$$GS_{in}/[GkB(T_A + T_n)] = [\text{input signal-to-noise ratio}]\,[T_A/(T_A + T_n)].$$

Relationship between Noise Factor and Noise Temperature

From equations (5.15) and (5.20) the noise generated within a circuit is $(F - 1)GkT_0B = GktT_0B$, or

$$t = F - 1. \hspace{4cm} (5.21)$$

The concept of noise temperature, instead of noise factor, is usually employed for expressing the noise picked up by an aerial and for low-noise circuits. For a low-noise circuit the use of noise factor to specify the noise performance will often lead to inconvenient numbers; consider, for example, a noise factor of 1.068 97 is a noise temperature of 20 K. Table 5.1 shows equivalent noise factor and noise temperature values.

System Noise Temperature

The system noise temperature T_S of a radio receiving system is the

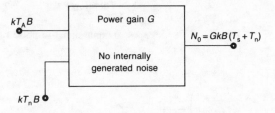

Fig. 5.10 Noise temperature.

Table 5.1

Noise factor (dB)	0.29	0.56	0.82	1.06	1.29	2.28	3.00
Noise temperature (K)	20	40	60	80	100	200	290

sum of the aerial noise temperature T_A and the overall noise temperature of the system, i.e. $T_S = T_A + T_{ov}$. The overall noise temperature T_{ov} is easily obtained by combining equations (5.18) and (5.21). Subtracting 1 from both sides of equation (5.18) gives

$$t_{ov} = t_1 + \frac{t_2}{G_1} + \frac{t_2}{G_1 G_2} \tag{5.22}$$

$$\text{or} \quad T_{ov} = T_1 + \frac{T_2}{G_1} + \frac{T_2}{G_1 G_2}. \tag{5.23}$$

The system output noise power N_0 is then given by equation (5.24)

$$N_0 = kT_S BG. \tag{5.24}$$

This method of calculating the output noise power from a radio system is particularly useful when dealing with a system in which the aerial noise temperature is not equal to 290 K.

Example 5.5

Calculate (a) the system noise temperature, (b) the output noise power, and (c) the output signal-to-noise ratio of the system shown by Fig. 5.11. The available signal power from the aerial is 3 pW, and the bandwidth is 10 MHz.

Solution

(a) Converting the decibel figures into ratios, 1 dB = 1.26, 20 dB = 100 and −10 dB = 0.1. The feeder has a noise factor of 1.26 and so its noise temperature is 75.4 K. The system noise temperature is

$$T_S = 40 + 75.4 + \frac{50}{1/1.26} + \frac{630}{100/1.26} + \frac{500}{10^4/1.26} = 186.4 \text{ K.}$$

$$(Ans.)$$

(b) $N_0 = GkT_S B = \dfrac{1}{1.26} \times 100 \times 100 \times 0.1 \times 1.38 \times 10^{-23} \times$

$186.4 \times 10^7 = 20.4$ pW. *(Ans.)*

(c) Output signal power $= 1000 \times 1/1.26 \times 3 \times 10^{-12} = 2.38\ \mu$W. Therefore

$$\text{output signal-to-noise ratio} = \frac{2.38 \times 10^{-6}}{20.4 \times 10^{-12}}$$

$$= 51 \text{ dB.} \quad (Ans.)$$

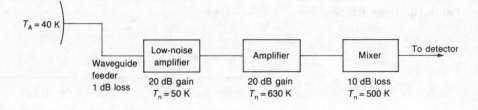

Fig. 5.11

6 Principles of Aerials

Transmitting and receiving aerials provide the link between the radio transmitter and receiver, and the propagation path via the atmosphere. All types of aerial are able both to transmit and receive radio-frequency energy and for each purpose will have the same gain and radiation pattern. It is customary to consider the operation of most aerials in their transmitting mode and to use the principle of reciprocity to obtain the receiving characteristics if, and when, required. Some aerials, such as those employed in broadcast systems, are required to transmit energy equally well in all directions in the horizontal plane. Other aerials are required to concentrate their radiation in one direction, these must have a directive radiation pattern to achieve the maximum gain in the wanted direction. Thus the directivity of an aerial, expressed graphically by means of its radiation pattern, is an important parameter.

For the maximum efficiency in the radiation of energy an aerial should be of resonant length; this means that its electrical length should be one-half a wavelength ($\lambda/2$). This requirement can be satisfied at frequencies in the h.f., v.h.f., u.h.f. and s.h.f. bands, and it is possible, although more difficult, at some frequencies in the m.f. band. It is not possible to obtain a $\lambda/2$ aerial at frequencies in the low-, and very-low-frequency bands because of the enormous physical structures that would be necessary. Transmitting aerials for use in the v.l.f., l.f. and m.f. bands are always vertical structures that are mounted upon the earth. The aerials are fed between the base of the aerial and the earth; ground reflections then make the aerial appear to be up to twice its physical height. An aerial whose electrical length is less than one-quarter wavelength, including the ground-plane image, is said to be an electrically short (or small) aerial.

Radiation from an Aerial

When a radio-frequency current flows in a conductor a magnetic field is set up around that conductor, the magnitude of which is directly proportional to the instantaneous value of the current. As the current varies with time the magnetic field will change as well. The changing magnetic field produces a changing electric field, not only in the vicinity of the magnetic field but also in the region surrounding it. Similarly, the changing electric field produces a further changing magnetic field in the next surrounding region, and so on. Radio-frequency energy is, in this way, propagated away from the conductor in the form of an electromagnetic wave. Figure 4.2 shows an electromagnetic wave in which the electric and magnetic fields vary sinusoidally with time. The two fields are always in phase quadrature and are mutually at right angles to their direction of propagation.

The plane that contains the electric field and the direction of propagation is known as the *polarization* of the wave. If, for example, the electric field is in the vertical plane the wave is said to be vertically polarized.

The electromagnetic wave propagates through the atmosphere with a velocity equal to the velocity of light c, which is generally (for radio work) taken as being equal to 3×10^8 m/s. At all times during its travel the two fields of the electromagnetic wave vary in phase with one another. The ratio of the amplitude of the electric field strength to the amplitude of the magnetic field strength is a constant that is known as the *impedance of free space*.

$$\text{Impedance of free space} = \frac{E \text{ (V/m)}}{H \text{ (A T/m)}} = 120\pi \; \Omega. \quad (6.1)$$

The power density P_d of the electromagnetic wave is equal to the product of the r.m.s. values of the electric and the magnetic field strengths, i.e.

$$P_d = EH \text{ W/m} = \frac{E^2}{120\pi} \text{ W/m}. \quad (6.2)$$

Since both the electric field and the magnetic field lie in a plane that is at right angles to the direction of propagation a radio wave is a transverse electromagnetic or TEM wave.

Isotropic Radiator

An *isotropic radiator* is a theoretical aerial, that cannot be realized in practice, which is able to radiate energy equally well in all the possible planes. The radiated energy will therefore have a spherical wavefront with its power spread uniformly over the surface of a sphere. The surface area of a sphere of radius r is $4\pi r^2$ and so the power density P_d of the radiated energy is

$$P_d = \frac{P_t}{4\pi D^2} \text{ W/m}, \quad (6.3)$$

where P_t is the transmitted power and D is the distance from the radiator. Clearly, the power density is inversely proportional to the square of the distance from the radiator. The power density is also, from equation (6.2), equal to $E^2/120\pi$, and equating the two relationships gives the electric field strength at distance D from the isotropic radiator as

$$E = \frac{\sqrt{(30P_t)}}{D} \text{ V/m}. \quad (6.4)$$

This, of course, is inversely proportional to distance. If the transmitted power is given in kilowatts and the distance D in kilometres, then $E = (173\sqrt{P_t})/D$ mV/m.

Although the isotropic radiator is not a practical aerial it is commonly employed as a reference against which other, practical, aerials can be compared.

Current Element

A current element consists of a current of uniform amplitude I amperes which flows in a very short length dl of conductor. It is sometimes known as a Hertzian dipole and it also cannot be practically realized. The current element is an extremely useful concept because the results obtained from a study of it can be applied to practical cases if another concept, that of the effective length of an aerial, is employed. Many practical aerials may be conveniently regarded as consisting of a large number of current elements in cascade.

When a radio-frequency current flows in a very short electrical length dl of conductor (see Fig. 6.1) to form a current element, the magnetic field that is set up will have a field strength given by

$$H = \frac{I \, dl \sin \theta}{4 \pi} \left[\frac{\omega}{cD} \cos \omega \left(t - \frac{D}{c} \right) \right.$$

$$\left. + \frac{\sin \omega}{D^2} \left(t - \frac{D}{c} \right) \right] \text{A T/m}, \tag{6.5}$$

where c is the velocity of light, D is the distance, in metres from the current element, and θ is the angle shown in Fig. 6.1.

Unlike the isotropic radiator, the current element does not radiate energy equally well in all directions but, instead, it produces a field that is proportional to $\sin \theta$. Equation (6.5) indicates that the magnetic field has two components which are known, respectively, as the *radiation field* and the *induction field*. The induction field is inversely proportional to the square of the distance from the current element and, therefore, its amplitude rapidly falls to a negligible value. The induction field represents the energy that is not radiated away from the current element. The magnitude of the radiation field is directly proportional to the frequency of the current flowing in the element and inversely proportional to distance. It therefore represents the energy that is radiated away from the current element.

The induction and radiation fields are of equal magnitude at the distance from the current element at which

$$\frac{\omega}{cD} = \frac{1}{D^2}$$

$$\text{or} \quad D = \frac{c}{\omega} = \frac{\lambda}{2\pi}.$$

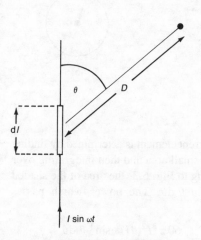

Fig. 6.1 Current element.

The distance $D = \lambda/2\pi$ marks the boundary between the *near field* and the *far field*. At distances greater than $\lambda/2\pi$ the induction field makes an insignificant contribution to the total field and so it is neglected.

The magnetic radiation field has an r.m.s. value of

$$H = \frac{I \, \mathrm{d}l\omega \sin \theta}{4\pi c D} = \frac{I \, \mathrm{d}l \sin \theta}{2\lambda D}. \tag{6.6}$$

An expression for the r.m.s. electric field strength is obtained by multiplying H by 120π to give

$$E = \frac{60\pi I \, \mathrm{d}l \sin \theta}{\lambda D}. \tag{6.7}$$

If $I \, \mathrm{d}l = 1$ mA and $D = 1$ km then the field strength is equal to 188 mV/m.

In the meridian plane (the plane that contains the current element) the field strength varies with $\sin \theta$ and the radiation pattern is as shown by Fig. 6.2(a). In the equatorial plane, where $\theta = 90°$ and hence $\sin \theta = 1$, there is equal radiation in all directions and so the radiation pattern is a circle (see Fig. 6.2(b)).

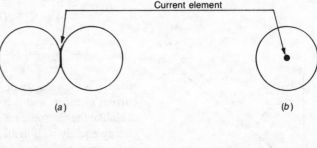

Current element

Fig. 6.2 Radiation patterns of a current element, (a) meridian plane and (b) equatorial plane.

(a) (b)

Radiated Power

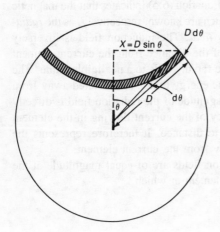

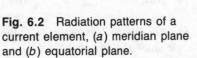

The total power radiated by a current element is determined by finding the power that passes through a small area and then integrating over the surface of a sphere. Referring to Fig. 6.3, the area of the shaded zone is $2\pi x D \, \mathrm{d}\theta = 2\pi D^2 \sin \theta \, \mathrm{d}\theta$. The power which passes through this zone is

$$\frac{E^2}{120\pi} \times 2\pi D^2 \sin \theta \, \mathrm{d}\theta = \frac{60\pi^2 I^2 (\mathrm{d}l)^2 \sin^3 \theta \, \mathrm{d}\theta}{\lambda^2} \text{ W}.$$

The total power P_t radiated over the surface of the sphere is

$$P_t = 2 \int_0^{\pi/2} \left[\frac{60\pi^2 I^2 (\mathrm{d}l)^2 \sin^3 \theta}{\lambda^2} \right] \mathrm{d}\theta$$

$$= \frac{120\pi^2 I^2 (\mathrm{d}l)^2}{\lambda^2} \int_0^{\pi/2} \frac{(3 \sin \theta - \sin 3\theta)}{4} \, \mathrm{d}\theta$$

Fig. 6.3 Power radiated by a current element.

$$= \frac{120\pi^2 I^2 (\mathrm{d}l)^2}{4\lambda^2} \left[-3\cos\theta + \frac{\cos 3\theta}{3} \right]_0^{\pi/2}$$

$$= \frac{30\pi^2 I^2 (\mathrm{d}l)^2}{\lambda^2} [3 - \tfrac{1}{3}]$$

$$\text{or} \quad P = \frac{80\pi^2 I^2 (\mathrm{d}l)^2}{\lambda^2} \text{ W.} \tag{6.8}$$

Expressing both equations (6.7) and (6.8) in terms of I^2 and then equating them gives

$$I^2 = \frac{P_t \lambda^2}{80\pi^2 (\mathrm{d}l)^2} = \frac{E^2 \lambda^2 D^2}{60\pi^2 (\mathrm{d}l)^2}$$

$$\text{or} \quad E = \frac{\sqrt{(45 P_t)}}{D}. \tag{6.9}$$

Radiation Resistance

The total power P_t radiated from an aerial may be considered to be equal to the product of the square of the input current to the aerial and a non-physical resistance R_r, i.e. $P_t = I^2 R_r$. This concept gives the *radiation resistance* of a current element as

$$R_r = \frac{80\pi (\mathrm{d}l)^2}{\lambda^2}. \tag{6.10}$$

Effective Length of an Aerial

The equations previously obtained for the electric field produced by, and the power radiated from, a current element cannot be directly applied to a practical application. They can, however, be usefully employed if another concept, that of the effective length of an aerial, is introduced.

The *effective length* l_{eff} of an aerial is that length which, if it carried a uniform current at the same amplitude as the input current I to the aerial, would produce the same field strength at a given point in the equatorial plane of the aerial. This means that the product of the physical length of the aerial and the mean current flowing in the aerial must be equal to the product of the effective length and the assumed uniform current. Thus

$$l_{\mathrm{phy}} I_{\mathrm{mean}} = l_{\mathrm{eff}} I$$

$$\text{or} \quad l_{\mathrm{eff}} = \frac{l_{\mathrm{phy}} I_{\mathrm{mean}}}{I} \tag{6.11}$$

$$\text{or} \quad l_{\mathrm{eff}} = \frac{l_{\mathrm{phy}}}{I} \left[\frac{1}{l_{\mathrm{phy}}} \int_0^{l_{\mathrm{phy}}} I(y)\, \mathrm{d}y \right] \tag{6.12}$$

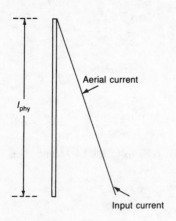

Fig. 6.4 Current distribution on an electrically short aerial.

or $\quad l_{\text{eff}} = \dfrac{1}{I} \displaystyle\int_0^{l_{\text{phy}}} I(y)\, \mathrm{d}y.$ (6.13)

Alternatively, the effective length of a receive aerial may be defined as $l_{\text{eff}} = V_{\text{oc}}/E$, where V_{oc} is the voltage that appears at the open-circuited terminals of the aerial when it is situated in an electric field of strength E V/m.

For an electrically short aerial, such as that shown in Fig. 6.4, the current distribution can be assumed to vary linearly from its maximum value I at the input terminals to zero at the top of the aerial. Clearly, the mean value of the aerial current is $I/2$ and, from equation (6.11), $l_{\text{eff}} = l_{\text{phy}}/2$. Using equation (6.12)

$$l_{\text{eff}} = \frac{l_{\text{phy}}}{I}\left[\frac{1}{l_{\text{phy}}}\int_0^{l_{\text{phy}}}\frac{I}{l_{\text{phy}}}\,(l_{\text{phy}} - y)\,\mathrm{d}y\right]$$

$$= \frac{1}{l_{\text{phy}}}\int_0^{l_{\text{phy}}}(l_{\text{phy}} - y)\,\mathrm{d}y = \frac{1}{l_{\text{phy}}}\left[l_{\text{phy}}y - \frac{y^2}{2}\right]_0^{l_{\text{phy}}}$$

$$= \frac{1}{l_{\text{phy}}}\left[l_{\text{phy}}^2 - \frac{l_{\text{phy}}^2}{2}\right] = \frac{l_{\text{phy}}}{2}.$$

Using equation (6.13)

$$l_{\text{eff}} = \frac{1}{I}\int_0^{l_{\text{phy}}}\frac{I}{l_{\text{phy}}}\,(l_{\text{phy}} - y)\,\mathrm{d}y = \frac{l_{\text{phy}}}{2}.$$

The radiation resistance of such an aerial is given by

$$R_{\text{r}} = 80\pi^2\left(\frac{l_{\text{phy}}}{2\lambda}\right)^2 = 20\pi^2\left(\frac{l_{\text{phy}}}{\lambda}\right)^2\ \Omega.$$

For longer aerials the assumption of a linear current distribution is no longer valid and it is customary to assume the distribution to be sinusoidal.

The Monopole Aerial

A monopole aerial is one which is mounted vertically upon the surface of the earth and which is fed between the base of the aerial and earth. This type of aerial is employed in the v.l.f., l.f. and m.f. bands; in the two lower bands the aerial must, of necessity, be electrically short, but in the m.f. band it is possible to employ aerials whose electrical length is $\lambda/4$ or even longer.

Figure 6.5 shows an aerial of physical height l_{phy} which is mounted vertically upon the earth. The aerial will radiate energy equally well in all directions in the horizontal plane but it will exhibit some *directivity* in the vertical plane. Some energy is directed upwards towards the sky whilst some other energy is radiated downwards towards the earth. The aerial site is chosen to ensure that the earth in the neighbourhood of the aerial is both flat and of high conductivity.

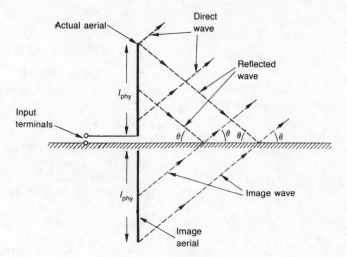

Fig. 6.5   Monopole aerial.

Therefore the waves radiated towards the earth are totally reflected with an angle of reflection equal to the angle of incidence. This makes the ground-reflected waves travel in the same direction as the direct waves. At a distant point P from the aerial energy is received by means of both the direct and the ground-reflected waves. The total field strength at this point is the phasor sum of the individual field strengths produced by each of the two waves. From the viewpoint of an observer at the point P it appears as though the ground-reflected wave has originated from an *image aerial* beneath the earth. This effect makes the aerial appear to be of *twice* its actual physical height and so doubles the effective length of the aerial. This means that an electrically short monopole aerial has an effective length equal to its physical length.

The electric field strength produced at the distant point P is hence

$$E = \frac{120\pi I \, l_{\text{eff}} \sin\theta}{\lambda D} \text{ V/m.} \tag{6.14}$$

The monopole aerial produces a similar radiation pattern, in both the horizontal and the vertical planes, as a balanced dipole of twice the physical length, that is situated in free space.

The power radiated by a monopole aerial is obtained by the use of equation (6.8) but since there is no electric field beneath the ground the actual power is only one-half of that predicted, i.e.

$$P_t = \frac{40\pi^2 I^2 (2l_{\text{eff}})^2}{\lambda^2} = 160\pi^2 \left(\frac{l_{\text{eff}}}{\lambda}\right)^2 I^2. \tag{6.15}$$

If both equation (6.14) and equation (6.15) are expressed in terms of I^2 and then equated

$$I^2 = \frac{P_t \lambda^2}{160\pi^2 l_{\text{eff}}^2} = \frac{E^2 \lambda^2 D^2}{(120\pi)^2 l_{\text{eff}}^2}$$

or $E = \dfrac{\sqrt{(90P_t)}}{D}.$ $\qquad$ (6.16)

If P_t is in kilowatts and D is in kilometres, then

$$E = \frac{\sqrt{(90 \times 10^3)}\sqrt{P_t}}{D} = \frac{300}{D}\sqrt{P_t} \text{ mV/m.} \qquad (6.17)$$

Example 6.1

A monopole aerial is 25 m high and it is supplied with a current of 100 A at 200 kHz. Assuming the current distribution on the aerial is linear calculate (*a*) the power radiated by the aerial, and (*b*) the field strength produced at ground level at a point 100 km distant.

Solution

$$\lambda = \frac{3 \times 10^8}{200 \times 10^3} = 1500 \text{ m.}$$

The electrical length of the aerial is $25/1500 = 0.017\lambda$.

(*a*) $P_t = 160\pi^2 \left(\dfrac{12.5}{1500}\right)^2 \times 10^4 = 1097 \text{ W.}$ (*Ans.*)

(*b*) From equation (6.14)

$$E = \frac{120\pi \times 100 \times 12.5}{1500 \times 100 \times 10^3} = 3.14 \text{ mV/m.} \qquad (Ans.)$$

Alternatively, using equation (6.16),

$$E = \frac{\sqrt{(90 \times 1097)}}{100} = 3.14 \text{ mV/m.} \qquad (Ans.)$$

As the electrical length of a monopole aerial is increased the assumption of a linear current distribution can no longer be made. The current distribution becomes an increasingly large part of a complete sinewave, and when the electrical length is one wavelength a complete cycle is described. If the length is less than $\lambda/4$ it is only necessary to determine the effective length of the monopole. If the length is $\lambda/4$ or more it will be necessary to consider the monopole to consist of the cascade connection of a large number of current elements. Each current element produces an electric field, due both to a direct and to a ground-reflected wave, given by equation (6.14), at a distant point P. The total field strength at the distant point is then the phasor sum of the field strengths due to each current element. This will result in the addition, or cancellation (perhaps complete), at different angles in the vertical plane; a number of examples are shown by Fig. 6.6. The vertical-plane radiation pattern changes only slowly with increase in the aerial height for heights up to about $\lambda/4$, and then, for heights between $\lambda/4$ and $\lambda/2$, becomes somewhat flatter in shape. The maximum radiation at ground level occurs when the electrical length of the aerial is $5\lambda/8$, although there is then some unwanted skywards

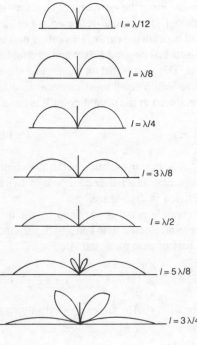

$l = \lambda/12$

$l = \lambda/8$

$l = \lambda/4$

$l = 3\lambda/8$

$l = \lambda/2$

$l = 5\lambda/8$

$l = 3\lambda/4$

Fig. 6.6 Radiation patterns of monopole aerials of different heights.

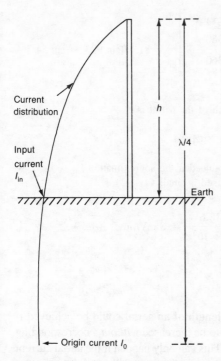

Current distribution

Input current I_{in}

Origin current I_0

Fig. 6.7 Current distribution on a monopole aerial.

radiation. If the electrical length is increased to more than $5\lambda/8$ the radiation of energy is concentrated in the direction of the sky.

The radiation patterns differ from that of a current element because the distances from each part of the monopole to the distant point P may differ by an appreciable part of a wavelength. This will not affect the amplitude of the individual field strengths but it does introduce phase differences.

Effective Length of a Short Monopole

When the electrical length of a monopole is short the current distribution on the aerial can be assumed to be a part of a cosinusoidal wave. As shown by Fig. 6.7 the current will be zero at the top of the aerial and increases to a maximum value I_0 at the origin which is at a distance $l = \lambda/4$ below the top of the aerial. Since the height of the aerial is less than $\lambda/4$ the origin is at a point situated $(l - h)$ below the surface of the earth. The current $I(y)$ at any point from the base of the aerial is given by

$$I(y) = I_0 \cos\left[\frac{2\pi y}{\lambda}\right], \qquad (6.18)$$

where y varies from $(l - h)$ to l.

Example 6.2

A monopole aerial is 36 m in height and is supplied with an input current of 30 A peak at a frequency of 833 kHz. Calculate (a) the effective height of the aerial, and (b) the field strength produced at ground level at a distance of 50 km.

Solution
At 833 kHz

$$\lambda = \frac{3 \times 10^8}{833 \times 10^3} = 360 \text{ m}.$$

Hence $\lambda/4 = 90$ m. From equation (6.18),

$$I_{in} = 30 = I_0 \cos\left[\frac{2\pi}{360} (90 - 36)\right] = I_0 \cos 54°.$$

Therefore, $I_0 = 30/(\cos 54°) = 51$ A. Hence

$$I(y) = 51 \cos\left[\frac{2\pi y}{360}\right]$$

and the mean value of the aerial current is

$$I_{mean} = 51 \times \frac{1}{90 - 54} \int_{54}^{90} \cos\left[\frac{2\pi y}{360}\right] dy$$

$$= 1.4167 \left[\frac{\sin \left(\frac{2\pi y}{360} \right)}{2\pi/360} \right]_{54}^{90} = 81.17[\sin 90° - \sin 54°]$$

$$= 15.5 \text{ A}.$$

(a) Hence the effective length of the aerial is

$$\frac{36 \times 15.5}{30} = 18.6 \text{ m}. \quad (Ans.)$$

(Note that the error introduced if the approximation $l_{eff} = l_{phy}/2 = 36/2 = 18$ m had been used is small.)

(b) From equation (6.14)

$$E = \frac{120\pi \times 30 \times 18.6}{\sqrt{2} \times 360 \times 50 \times 10^3} = 8 \text{ mV/m}. \quad (Ans.)$$

Top Loading

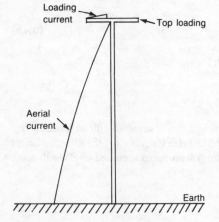

Fig. 6.8 Top loading of a monopole aerial.

An increase in the effective length of an aerial could be achieved if the mean aerial current were to be increased *without* a corresponding increase in the input current. This can only happen if the aerial current is prevented from falling to zero at the top of the aerial by the use of *top loading*. Top loading of a monopole aerial means that a horizontal system of conductors, which has a relatively large capacitance to earth, is fitted to the top of the aerial. The aerial current will flow in the top loading and so it will not fall to zero at the top of the radiating part of the aerial. The idea is illustrated by Fig. 6.8. The top loading makes little, or no, contribution to the total field strength produced at a distant point because it is only a small electrical distance from its under-earth image. This image is of the opposite polarity to the top-loading conductors and hence tends to cancel out any radiation from the top loading.

The effective length of a top-loaded monopole is calculated in a manner similar to that used before but the top loading must be taken into account. The total length of the aerial must now include the length of the top loading. Suppose, for example, that the aerial of Example 6.2 was fitted with a top-loading system in the form of a number of radially spaced conductors of length 5 m. Then the distance y would vary from $l = 90$ m to $(l - h) = 90 - 41 = 49$ m and $I_{mean} = 19.9$ A.

λ/4 Monopole

When the length of a monopole aerial is increased to λ/4, or more, the electrical distance between a current element in the actual aerial and the corresponding image current element leads to a phase difference between the field strengths produced at a distant point P. Consider Fig. 6.9, which shows a current element $I\,\mathrm{d}y$, distance y from the base of the aerial, which is radiating energy. The radiation from the image current element $I\,\mathrm{d}y$ has an extra distance $2y\cos\theta$ to travel before it reaches the point P. This extra distance produces a phase lag $\phi = (2\pi/\lambda)2y\cos\theta$; thus, referring to Fig. 6.10,

$$\cos\frac{\phi}{2} = \frac{E/2}{E_\mathrm{d}}$$

and $E = 2E_\mathrm{d}\cos\dfrac{\phi}{2}$.

Hence, the total field strength $\mathrm{d}E_\mathrm{T}$ due to the two current elements is

$$\mathrm{d}E_\mathrm{T} = 2E_\mathrm{d}\cos\frac{\phi}{2} = 2E_\mathrm{d}\cos\left[\frac{2\pi}{\lambda}y\cos\theta\right],$$

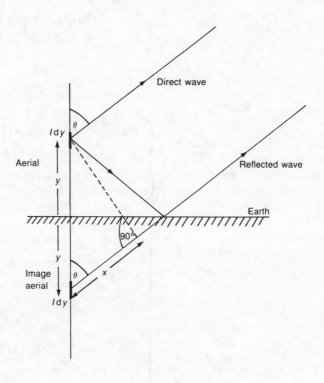

Fig. 6.9 Waves radiated by current elements in both the monopole aerial and its image.

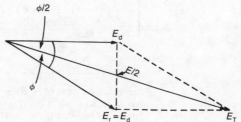

Fig. 6.10 Phasor diagram of the field strengths at a distant point.

where

$$E_d = \frac{60 \pi I \, dy \sin \theta}{\lambda D}.$$

The aerial current at distance y from the base of the aerial is

$$I(y) = I \cos \left[\frac{2\pi}{\lambda} y \right]$$

and hence

$$dE_T = \frac{120 \pi I \sin \theta}{\lambda D} \cos \left[\frac{2\pi}{\lambda} y \right] \cos \left[\frac{2\pi}{\lambda} y \cos \theta \right].$$

The total field strength produced at the point is then

$$E_T = \frac{120 \pi I \sin \theta}{\lambda D} \int_0^{\lambda/4} \left[\cos \left(\frac{2\pi}{\lambda} y \right) \cos \left(\frac{2\pi}{\lambda} y \cos \theta \right) \right] dy$$

$$= \frac{120 \pi I \sin \theta}{2 \lambda D} \int_0^{\lambda/4} \left\{ \cos \left[\frac{2\pi}{\lambda} y (1 + \cos \theta) \right] \right.$$

$$\left. + \cos \left[\frac{2\pi}{\lambda} y (1 - \cos \theta) \right] \right\} dy\dagger$$

$$= \frac{60 \pi I \sin \theta}{\lambda D} \left\{ \frac{\sin \left[\frac{2\pi}{\lambda} y (1 + \cos \theta) \right]}{\frac{2\pi}{\lambda} (1 + \cos \theta)} + \frac{\sin \left[\frac{2\pi}{\lambda} y (1 - \cos \theta) \right]}{\frac{2\pi}{\lambda} (1 - \cos \theta)} \right\}_0^{\lambda/4}$$

$$= \frac{60 \pi I \sin \theta}{\lambda D} \left[\frac{\sin \left(\frac{\pi}{2} + \frac{\pi}{2} \cos \theta \right)}{\frac{2\pi}{\lambda} (1 + \cos \theta)} + \frac{\sin \left(\frac{\pi}{2} - \frac{\pi}{2} \cos \theta \right)}{\frac{2\pi}{\lambda} (1 - \cos \theta)} \right]$$

$$= \frac{60 \pi I \sin \theta}{\lambda D} \left[\frac{\sin \frac{\pi}{2} \cos \left(\frac{\pi}{2} \cos \theta \right)}{\frac{2\pi}{\lambda} (1 + \cos \theta)} + \frac{\sin \frac{\pi}{2} \cos \left(\frac{\pi}{2} \cos \theta \right)}{\frac{2\pi}{\lambda} (1 - \cos \theta)} \right] \dagger\ddagger$$

$\dagger \cos A \cos B = \cos (A + B) + \cos (A - B)$
$\dagger \sin (A + B) = \sin A \cos B + \cos A \sin B$
$\ddagger \sin (-A) = -\sin A, \cos (-A) = \cos A$

$$= \frac{60\pi I \sin \theta}{\lambda D} \left[\frac{\cos\left(\frac{\pi}{2} \cos \theta\right)}{\frac{2\pi}{\lambda}(1 + \cos \theta)} + \frac{\cos\left(\frac{\pi}{2} \cos \theta\right)}{\frac{2\pi}{\lambda}(1 - \cos \theta)} \right]$$

$$= \frac{30I \sin \theta}{D} \left[\frac{\cos\left(\frac{\pi}{2} \cos \theta\right)}{1 + \cos \theta} + \frac{\cos\left(\frac{\pi}{2} \cos \theta\right)}{1 - \cos \theta} \right]$$

$$= \frac{30I \sin \theta}{D} \left[\frac{2 \cos\left(\frac{\pi}{2} \cos \theta\right)}{1 - \cos^2 \theta} \right]$$

$$\text{or}\quad E_T = \frac{60I}{D} \left[\frac{\cos\left(\frac{\pi}{2} \cos \theta\right)}{\sin \theta} \right]. \tag{6.19}$$

A similar analysis for a $\lambda/2$ monopole (using $i = I \sin\left(\frac{2\pi}{\lambda} y\right)$) results in

$$E_T = \frac{60I}{D} \left[\frac{\cos(\pi \cos \theta) + 1}{\sin \theta} \right]. \tag{6.20}$$

The λ/2 Dipole

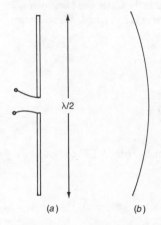

Fig. 6.11 (a) λ/2 dipole, (b) current distribution on a λ/2 dipole.

A dipole is a single conductor of length l that is centre fed as shown by Fig. 6.11(a). Nearly always the length is such that the dipole is resonant at the frequency of the input signal. This means that it is one-half wavelength long and has the current distribution shown in Fig. 6.11(b). In practice, the dipole length is often made slightly less than $\lambda/2$ in order to achieve a purely resistive input impedance. At frequencies in the h.f. band, and particularly in the v.h.f. and u.h.f. bands, the physical dimensions of the $\lambda/2$ dipole make it the basic element of many types of aerial array.

The electric field set up by a $\lambda/2$ dipole at a distance D is given by equation (6.19). When $\theta = 90°$, which defines the equatorial plane of the dipole, the radiation pattern is a circle (see Fig. 6.12(a)). In the plane of the aerial the radiation pattern is a figure-of-eight shape, as shown by Fig. 6.12(b). The *beamwidth* of the pattern, i.e. the angle subtended by the 3 dB points, is $129° - 51° = 78°$.

Effective Length

The effective length l_{eff} of a $\lambda/2$ dipole is, from equation (6.13),

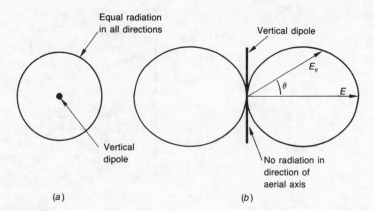

Fig. 6.12 Radiation patterns of a
λ/2 dipole: (*a*) equatorial plane,
(*b*) meridian plane.

(*a*)

(*b*)

$$l_{\text{eff}} = \frac{1}{I} \int_{-\lambda/4}^{\lambda/4} \left[I \cos \left(\frac{2\pi y}{\lambda} \right) \right] dy$$

$$= \frac{1}{I} \left[\frac{I}{2\pi/\lambda} \sin \left(\frac{2\pi}{\lambda} \right) y \right]_{-\lambda/4}^{\lambda/4} = \frac{\lambda}{\pi}. \qquad (6.21)$$

Radiation Resistance

If the effective length λ/π of a λ/2 dipole is substituted into equation (6.8), the value obtained for the radiation resistance will be 80 Ω. This figure is, however, incorrect. The correct value for the radiation resistance of a λ/2 dipole is 73.14 Ω.

Example 6.3

Calculate the electric field strength produced in the equatorial plane of a λ/2 dipole at a distance of 20 km if the input current is 3 A r.m.s. Also calculate the power density of the wave at this point.

Solution
From equation (6.20)

$$E = \frac{60 \times 3}{20} = 9 \text{ mV/m.} \qquad (Ans.)$$

$$\text{Power density} = \frac{E^2}{120\pi} = 215 \text{ nW/m.} \qquad (Ans.)$$

Alternatively, radiated power

$$P_t = I^2 R_r = 9 \times 73.14 = 658.3 \text{ W}$$

$$\text{and power density} = \frac{P_t G_t\dagger}{4\pi D^2} = \frac{658.3 \times 1.64}{4 \times 400 \times 10^6} = 215 \text{ nW/m.}$$

$$(Ans.)$$

†See p. 127.

Gain of an Aerial

The *gain* of an aerial indicates the extent to which the energy that it radiates is concentrated in a particular direction, *or* the extent to which the aerial receives signals better from one direction than from all others. The gain of an aerial is defined relative to a reference aerial which is usually either an isotropic radiator or a $\lambda/2$ dipole. When the gain is with reference to the isotropic radiator it is expressed in dBi. The gain of an aerial is the same whether the aerial is employed to transmit or to receive signals.

The gain of an aerial may be defined in two ways.

(*a*) The gain is the square of the ratio of the field strength produced at a point in the direction of maximum radiation, to the field strength produced at the same point by the reference aerial, both aerials radiating the same power.

(*b*) The gain is the ratio of the powers that the aerial, and the reference aerial, must radiate to set up the same field strength at a point in the direction of maximum radiation.

Current Element

From equation (6.4) and equation (6.9) the gain of a current element is

$$G = \left(\frac{\sqrt{(45 P_t)}}{D} \middle/ \frac{\sqrt{(30 P_t)}}{D} \right)^2 = 1.5 \quad \text{or} \quad 1.76 \text{ dBi}.$$

Monopole Aerial

From equations (6.4) and (6.16)

$$G = \left(\frac{\sqrt{(90 P_t)}}{D} \middle/ \frac{\sqrt{(30 P_t)}}{D} \right)^2 = 3 \quad \text{or} \quad 4.77 \text{ dBi}.$$

$\lambda/2$ Dipole

The power P_t radiated by a $\lambda/2$ dipole is equal to $73.14 I^2$ and substituting I^2 into equation (6.19) gives

$$E = \frac{60 \sqrt{(P_t/73.14)}}{D}.$$

Therefore, the gain of a $\lambda/2$ dipole is

$$G = \left(\frac{60 \sqrt{(P_t/73.14)}}{D} \middle/ \frac{\sqrt{(30 P_t)}}{D} \right)^2 = 1.64 \quad \text{or} \quad 2.15 \text{ dBi}.$$

Effective Radiated Power

The effective radiated power (e.r.p.) of an aerial is the power which an isotropic radiator would have to radiate to produce the same field strength at a point in the direction of maximum radiation. Numerically, e.r.p. is equal to the product of the total transmitted power P_t and the gain G_t of the aerial, i.e.

$$\text{e.r.p.} = G_t P_t.$$ (6.22)

Using this concept, equation (6.4) can be adapted to give the electric field strength produced by an aerial of gain G_t, i.e.

$$E = \frac{\sqrt{30 P_t G_t}}{D}.$$ (6.23)

Effective Aperture of an Aerial

The effective aperture A_e of an aerial is the imaginary cross-sectional area that would absorb the same power from an incident wave as does the aerial when it is matched to its load. If the incident radio wave has a power density of $P_d = E^2/120\pi$ W/m, then the power P_r received by the aerial will be equal to $P_d A_e$. The gain G of an aerial is directly proportional to its effective aperture; thus

$$G = \frac{A_e}{\text{effective aperture of isotropic radiator } A_{e(iso)}}.$$ (6.24)

$\lambda/2$ Dipole

The voltage V induced into a $\lambda/2$ dipole by an incident radio wave is the product of the electric field strength and the effective length of the dipole. Thus $V = E\lambda/\pi$ volts. The dipole has an input resistance that is equal to its radiation resistance of $73.14\ \Omega$, and for the maximum transfer of power to the load, the load resistance must also be equal to $73.14\ \Omega$. Then (see Fig. 6.13)

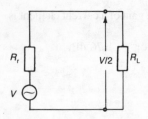

Fig. 6.13 Determination of the effective aperture of a $\lambda/2$ dipole.

$$\frac{E^2 A_e}{120\pi} = \frac{V^2}{4R_r} = \frac{E^2 \lambda^2}{4\pi^2 \times 73.14}$$

or $$A_e = \frac{30\lambda^2}{73.14\pi} = 0.13\lambda^2.$$ (6.25)

Current Element

Since the current flowing in the element is assumed to be of uniform amplitude the voltage V induced into the aerial by an incident radio wave is equal to $E\,\mathrm{d}l$. Therefore

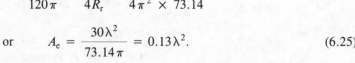

or $$A_e = \frac{3\lambda^2}{8\pi} = 0.12\lambda^2.$$ (6.26)

Isotropic Radiator

A current element has a gain of 1.5 relative to an isotropic radiator, and so

$$1.5 = \frac{A_{e(ce)}}{A_{e(iso)}} = \frac{3\lambda^2}{8\pi A_{e(iso)}}$$

or $\quad A_{e(iso)} = \dfrac{\lambda^2}{4\pi}.$ \hfill (6.27)

Power Received by an Aerial

The power P_r received by an aerial when it is sited in a field of strength E V/m is

$$P_r = P_d A_e = \frac{E^2\lambda^2 G_r}{120\pi(4\pi)} = \frac{G_r}{30}\left(\frac{E\lambda}{4\pi}\right)^2.$$

From equation (6.23),

$$E = \frac{\sqrt{(30 P_t G_t)}}{D}$$

and substituting this gives

$$P_r = G_t G_r P_t \left(\frac{\lambda}{4\pi D}\right)^2. \hfill (6.28)$$

Example 6.4

The transmit aerial of a 600 MHz radio link has a gain of 26 dB. The signal is received by an identical aerial that is 30 km away. Calculate the overall loss of the radio link.

Solution
From equation (6.28)

$$P_r/P_t = 398^2 \times \left(\frac{3 \times 10^8}{4\pi \times 30 \times 10^3 \times 600 \times 10^6}\right)^2$$

$$= 2.79 \times 10^{-7}.$$

Therefore, the link loss = 65.5 dB.   (*Ans.*)

Example 6.5

A $\lambda/4$ monopole aerial is supplied with a current of 30 A r.m.s. at 3 MHz. Calculate (*a*) the field strength produced at ground level at a point 50 km distant, and (*b*) the power received by an aerial of 10 dBi gain situated at that point.

Solution

(*a*) $\quad \lambda = \dfrac{3 \times 10^8}{3 \times 10^6} = 100$ m

and hence $\lambda/4 = 25$ m. Since the monopole aerial is $\lambda/4$ long the input current is at the origin and so $I = 30\cos(2\pi y/100)$. The mean aerial current is

$$I_{mean} = \frac{1}{25} \int_0^{25} 30 \cos\left(\frac{2\pi y}{100}\right) dy = \frac{30}{25}\left[\frac{\sin}{2\pi/100}\left(\frac{2\pi y}{100}\right)\right]_0^{25}$$

$$= \frac{1.2 \times 100}{2\pi} \times \sin\left(\frac{50\pi}{100}\right) = 19.1 \text{ A.}$$

Therefore

$$l_{eff} = \frac{19.1 \times 25}{30} = 15.9 \text{ m.}$$

Substituting into equation (6.14)

$$E = \frac{120\pi \times 30 \times 15.9}{100 \times 50 \times 10^3} = 36 \text{ mV/m.} \quad (Ans.)$$

(b) The receive aerial has a gain of 10 dBi = 10 times, hence its effective aperture is

$$A_e = 10 \times A_{e(iso)} = \frac{10\lambda^2}{4\pi} = \frac{10 \times 10^4}{4\pi} = 7958 \text{ m}^2.$$

The power density at the distant point P is $P_d = E^2/120\pi = 3.44 \times 10^{-6}$ W/m, and therefore the received power is

$$P_r = A_e P_d = 7958 \times 3.44 \times 10^{-6} = 27.36 \text{ mW.} \quad (Ans.)$$

Long-wire Radiator

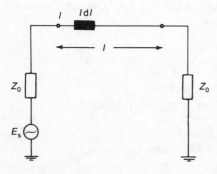

Fig. 6.14   Long-wire radiator.

Figure 6.14 shows the basic arrangement of a long-wire radiator. Essentially, it consists of a conductor, several wavelengths long at the frequencies of operation, which together with the earth forms a transmission line of characteristic impedance Z_0. At its sending end a radio-frequency source of e.m.f. E_S and impedance Z_0 supplies a current $I_S = E_S/2Z_0$. The attenuation of the line is negligibly small and so this current flows, with unchanged amplitude but varying phase, to the matched load. The length of line l can be considered to consist of the cascade connection of a large number of current elements. Each current element will radiate energy, the amplitude of which is at its maximum value in the equatorial plane and zero along the axis of the line. The total field strength produced at a distant point P is the phasor sum of the field strengths produced by each of the individual current elements.

Consider Fig. 6.15. The current at a distance x from the sending end of the line is $I_x = I_S e^{-j\beta x}$ and so it lags the sending-end current I by angle $\beta x = 2\pi x/\lambda$ radians. The current element $I_x \, dx$ is closer to the distant point P by $x \cos\theta$ metres and so the field strength dE produced by element $I_x \, dx$ lags the field strength due to the current element at the beginning of the line by angle $(2\pi x/\lambda)(1 - \cos\theta)$ radians. Now,

$$dE = \frac{j60\pi I \, dx \sin\theta}{\lambda D}$$

and hence the total field at P is

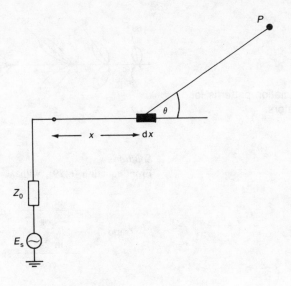

Fig. 6.15 Calculation of field strength.

$$E = \int_0^l \left[\left(\frac{j60\pi \sin \theta}{\lambda D} \right) I_S e^{-(j\pi x/\lambda)(1 - \cos \theta)} \right] dx$$

$$= \frac{j60\pi I_S \sin \theta}{\lambda D} \left[\frac{e^{-(j2\pi x/\lambda)(1 - \cos \theta)}}{(j2\pi x/\lambda)(1 - \cos \theta)} \right]_0^l$$

$$= \frac{30 I_S}{D} \frac{\sin \theta}{1 - \cos \theta} \left[e^{-(j2\pi l/\lambda)(1 - \cos \theta)} - 1 \right]$$

or $$E = \left[\frac{60 I_S}{D} \frac{\sin \theta}{1 - \cos \theta} \right] \left[\sin \left(\frac{\pi l}{\lambda} (1 - \cos \theta) \right) \right].† \quad (6.29)$$

The main lobes of the radiation pattern of a long-wire radiator for five different lengths are shown in Fig. 6.16. As the length of the wire increases the number of sidelobes increases also but these are not shown. When the length of the wire is between 4λ and 8λ there is a linear relationship between the angle of the main lobe and frequency. The long-wire radiator can be used as an aerial in its own right but, more often, it is a part of a rhombic aerial.

Example 6.6

Calculate the ratio of the voltages induced into a long-wire radiator of 100 m length if the incident radio wave is at 15 MHz and the angle of incidence is (a) 17°, and (b) 24°.

---

†$e^{-jx} = \cos x - j \sin x$, $|e^{-jx} - 1| = \sqrt{[(\cos x - 1)^2 + \sin^2 x]}$
$$= \sqrt{\left[4 \sin^2 \left(\frac{x}{2} \right) \right]} = 2 \sin \left(\frac{x}{2} \right).$$

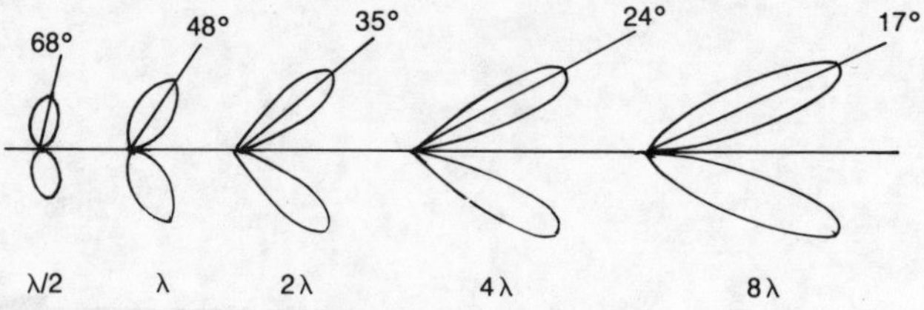

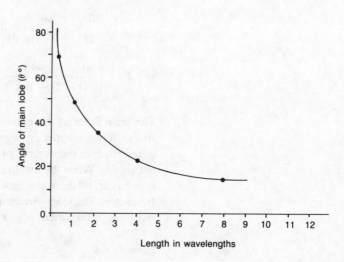

Fig. 6.16 Radiation patterns for long-wire radiators.

Solution

From equation (6.29), with $\lambda = 20$ m,

$$\text{ratio} = \frac{\dfrac{\sin 17°}{1 - \cos 17°} \sin \left[\dfrac{\pi \times 100}{20} (1 - \cos 17°) \right]}{\dfrac{\sin 24°}{1 - \cos 24°} \sin \left[\dfrac{\pi \times 100}{20} (1 - \cos 24°) \right]}$$

$$= 0.92. \quad (Ans.)$$

Values of the lobe angle θ for different values of wire length, in wavelengths, can be obtained from equation (6.29) and are shown plotted in Fig. 6.17. It can be seen that the lobe angle varies, more or less linearly, from 24° at $l = 4\lambda$ to 17° at $l = 8\lambda$.

Fig. 6.17 Main lobe angle; variation with wavelength.

7 Aerials

For radiocommunication between two points, high-gain, directive aerials are required for both transmission and reception. In the h.f. band the rhombic aerial has been widely employed for many years but it has now been superseded, for all but a few special applications, by the log-periodic aerial. The log-periodic aerial can provide an equal, or superior, performance in terms of both gain and bandwidth for a smaller physical size and, hence, cost. All h.f. radio links employ sky-wave propagation via the ionosphere, and for the best results the aerial must be selected to suit the propagation conditions pertaining at a given time. In particular, for reliable communications to be maintained, the system needs to be flexible in its usage of frequency and this means that both the transmitting aerial and the receiving aerial must be wideband. Nowadays, the selection of frequencies, elevation angle, etc., is made easier and quicker by the use of computer control.

Omnidirectional h.f. transmitting aerials are required for some purposes, and in the past were mainly vertical whip aerials. These are of fairly low gain and of inadequate performance at high elevation angles, and modern systems probably use a vertical stack of loop aerials fed in a log-periodic manner.

In the v.h.f. and u.h.f. bands the dimensions of a $\lambda/2$ dipole are small enough for linear arrays of dipoles to be the predominant type of aerial. Alternatives that are also often employed are the corner reflector, the Yagi, and the log-periodic aerials. For omnidirectional radiation of energy a vertically stacked sleeve dipole is often used. Mobile radio normally uses either the whip aerial or the helical aerial.

In the s.h.f. band high-gain parabolic dish aerials are the most common types, although horn aerials are also employed.

Arrays of Driven Dipoles

A $\lambda/2$ dipole has a gain of 2.16 dBi, a circular equatorial-plane radiation pattern, and a figure-of-eight radiation pattern in the meridian plane. For many applications this gain and directivity is inadequate and then two, or more, dipoles can be used in an array. A variety of radiation patterns can be obtained by varying one, or more, of: (a) the number of dipoles used, (b) the spacing between the dipoles, and (c) the amplitudes and relative phases of the dipole currents. The radiation pattern and the gain of an aerial may be affected by mutual impedances between the dipoles in an array, but this factor will not be considered until page 149. In the past arrays of dipoles were often employed as h.f. aerials, but today their main applications are in the v.h.f. and the u.h.f. bands.

Two-dipole Array

Vertical Dipoles

Horizontal-plane Radiation Pattern

Figure 7.1 shows two vertical $\lambda/2$ dipoles mounted in the same horizontal plane and spaced apart by a distance of d metres. If the dipoles are supplied with equal-amplitude, in-phase currents, both dipoles will radiate energy equally well in all directions in the horizontal plane, and the total field strength at any point in this plane will be the phasor sum of the individual field strengths produced by each aerial. If the distance to a distant point P is very much larger than the dipole spacing d these individual field strengths will be of equal amplitude. They will not usually, however, be in phase with one another. The point 0 is equidistant from both dipoles and so at this point the two fields are in phase and so will simply add. In general, the energy radiated by dipole A must travel a further distance $d \cos \theta$ to reach the point P. Therefore, the field strength due to dipole A will lag the field strength due to dipole B by an angle

$$\phi = \frac{2\pi d}{\lambda} \cos \theta \text{ radians.}$$

The phasor diagram of the field strengths at the point P is shown by Fig. 7.2. From this figure, the total field strength E_T is equal to

$$E_T = 2E \cos \frac{\phi}{2} = 2E \cos \left(\frac{\pi d \cos \theta}{\lambda} \right). \tag{7.1}$$

If either $\lambda/2$ dipole on its own was supplied with a power of P watts

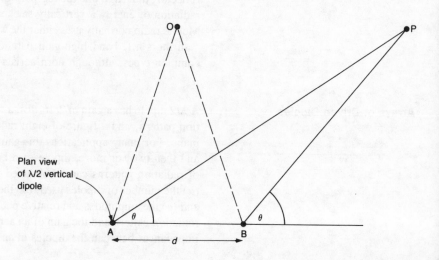

Plan view of $\lambda/2$ vertical dipole

Fig. 7.1 Two-dipole array.

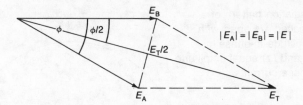

Fig. 7.2 Phasor diagram of the field strengths at a distant point.

the field strength produced at the point P would be $E_0 = 60I/D$. If the same power is supplied to the two-dipole array each dipole will receive a power of $P/2$ watts and so will produce a field strength of $E_0/\sqrt{2}$ at point P. Hence, the total field strength at the point P is

$$E_T = 2 \frac{E_0}{\sqrt{2}} \cos\left(\frac{\pi d \cos\theta}{\lambda}\right) = \sqrt{2}E_0 \cos\left(\frac{\pi d \cos\theta}{\lambda}\right).$$
(7.2)

The gain of the array, relative to a $\lambda/2$ dipole, is $20 \log_{10}\sqrt{2} = 3$ dB. The *array factor* is the term $\sqrt{2} \cos[(\pi d \cos\theta)/\lambda]$.

Example 7.1

Calculate and plot the horizontal radiation pattern of two vertical $\lambda/2$ dipoles which are $\lambda/4$ apart and supplied with equal-amplitude, in-phase currents.

Solution

$$\phi = \frac{2\pi}{\lambda} \frac{\lambda}{4} \cos\theta = \frac{\pi}{2} \cos\theta.$$

Hence the array factor is $\sqrt{2} \cos[\pi/4 \cos\theta]$. It is usually sufficient to calculate the resultant field strength at 30° intervals over the range 0° to 180°, see Table 7.1.

Using the figures given in the final row of Table 7.1 the radiation pattern for the array has been plotted and it is shown by Fig. 7.3(a).

When the currents in the two dipoles are of equal amplitude but there is a phase difference α between them the radiation pattern will be altered. The angle α must be added to ϕ to obtain a new phase difference $\psi = \phi - \alpha$ radians if the phase of I_A leads I_B, or $\psi = \phi + \alpha$ if the phase of I_B leads I_A. Very often, the value of α is made equal to the distance between the dipoles. If, for example, the dipole spacing is $\lambda/4$, α will be equal to 90°.

Table 7.1

θ	0°	30°	60°	90°	120°	150°	180°
$\cos\theta$	1	0.866	0.5	0	−0.5	−0.866	−1
45° $\cos\theta$	45°	39°	22.5°	0°	−22.5°	−39°	−45°
$\sqrt{2} \cos(45°\cos\theta)$	1	1.1	1.3	1.41	1.3	1.1	1

Fig. 7.3 Radiation pattern of a
two-dipole spaced λ/4 apart with
(a) equal-amplitude in-phase
currents, (b) and (c) equal-amplitude
90° out-of-phase currents.

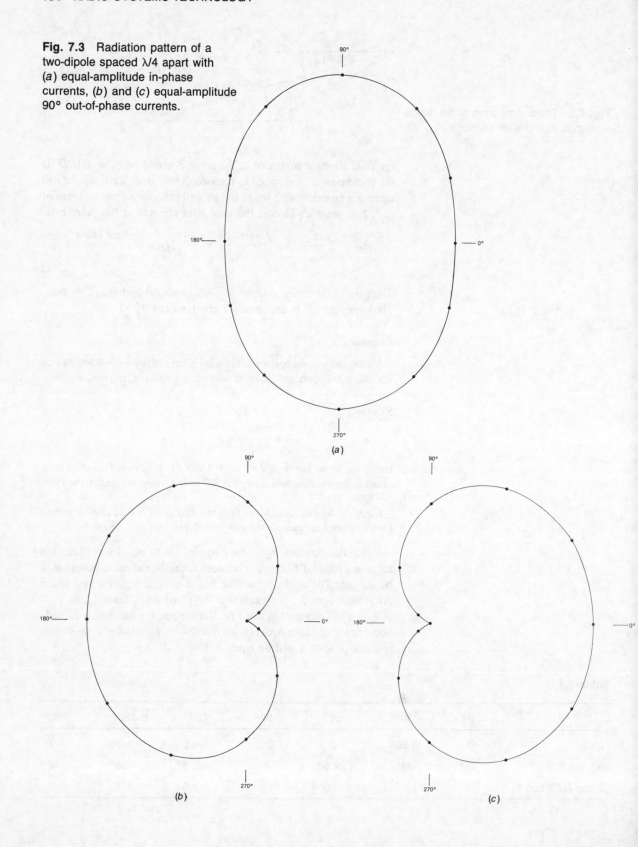

Example 7.2

Calculate and plot the horizontal plane radiation pattern for two vertical $\lambda/2$ dipoles that are $\lambda/4$ apart and are fed with equal-amplitude currents that are 90° out of phase.

Solution

If I_B leads I_A,

$$\psi = \frac{2\pi}{\lambda}\frac{\lambda}{4}\cos\theta + 90° = 90°(1 + \cos\theta)$$

and hence the array factor is $\sqrt{2}\cos[45°(1 + \cos\theta)]$. The results are tabulated in Table 7.2. The radiation pattern is shown in Fig. 7.3(b).

Conversely, if I_A leads I_B,

$$\psi = 90°(\cos\theta - 1)$$

and the array factor is equal to $\sqrt{2}\cos[45°(\cos\theta - 1)]$. The results are presented in Table 7.3. This radiation pattern is plotted in Fig. 7.3(c).

The currents supplied to the two dipoles may not always be of equal amplitude; Fig. 7.4(a) shows the phasor diagram of the field strengths at the distant point P when $|I_B| > |I_A|$. Taking E_B as the reference phasor and resolving E_A into its horizontal and vertical components gives the phasor diagram shown in Fig. 7.4(b). From this diagram the total field strength E_T is

$$E_T = \sqrt{[(E_B + E_A\cos\psi)^2 + E_A^2\sin^2\psi]}$$
$$= \sqrt{(E_B^2 + 2E_AE_B\cos\psi + E_A^2\cos^2\psi + E_A^2\sin^2\psi)}$$
$$\text{or}\quad E_T = \sqrt{(E_A^2 + E_B^2 + 2E_AE_B\cos\psi)}. \tag{7.3}$$

Table 7.2

θ	0°	30°	60°	90°	120°	150°	180°
$1 + \cos\theta$	2	1.866	1.5	1	0.5	0.134	0
$45°(1 + \cos\theta)$	90°	84°	67.5°	45°	22.5°	6°	0°
Array factor	0	0.15	0.54	1.0	1.31	1.41	1.41

Table 7.3

θ	0°	30°	60°	90°	120°	150°	180°
$\cos\theta - 1$	0	−0.134	−0.5	−1.0	−1.5	−1.866	−2
$45°(\cos\theta - 1)$	0°	−6°	−22.5°	−45°	−67.5°	−84°	−90°
Array factor	1.41	1.41	1.31	1.0	0.54	0.15	0

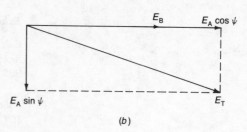

Fig. 7.4 (a) Phasor diagram of unequal-amplitude field strengths, (b) simplified version of (a).

(a)

(b)

Note that if $E_A = E_B = E$

$$E_T = \sqrt{[2E^2(1 + \cos \psi)]} = \sqrt{2}E\sqrt{(1 + \cos \psi)}$$
$$= \sqrt{2}E\sqrt{(2 \cos^2 \psi/2)} = 2E \cos \psi/2,$$

as before.

Vertical-plane Radiation Pattern

The vertical-plane radiation pattern of two vertical $\lambda/2$ dipoles mounted in the same horizontal plane is obtained by multiplying the array factor by equation (6.19), i.e.

$$E_T = \sqrt{2}E \cos \psi/2 \; \frac{\cos (\pi/2 \cos \theta)}{\sin \theta}. \tag{7.4}$$

Co-linear Vertical Dipoles

Two co-linear vertical dipoles (see Fig. 7.5) have a circular radiation pattern in the horizontal plane since, quite clearly, the energy radiated from each dipole does not combine with that from the other. In the vertical plane the radiation pattern is given by the product of the array factor and the dipole's meridian-plane pattern, i.e. by equation (7.4).

Dipole in Front of a Reflecting Plane

If a $\lambda/2$ dipole is mounted in front of a reflecting plane the image of the dipole will act as though it were the second dipole in a two-dipole array. The polarity of the image dipole depends upon the orientation of the physical dipole; Fig. 7.6 shows the two most common cases. When, Fig. 7.6(a), the dipole is mounted parallel to the reflecting plane the image dipole has the opposite polarity. Conversely, Fig. 7.6(b), if the dipole is mounted normal to the reflecting plane, both the dipole and its image have the same polarity. The apparent spacing d between the two 'dipoles' is twice the distance s between the dipole and the reflecting plane.

One application of this principle is the *corner reflector*. A $\lambda/2$ dipole is placed in a position parallel to the intersection of two reflecting planes as shown by Fig. 7.7(a). The reflecting planes are made of

Fig. 7.5 Co-linear dipoles.

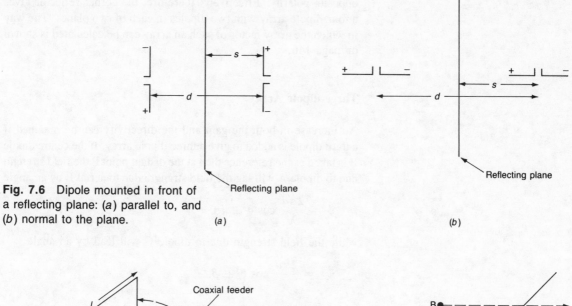

Fig. 7.6 Dipole mounted in front of a reflecting plane: (*a*) parallel to, and (*b*) normal to the plane.

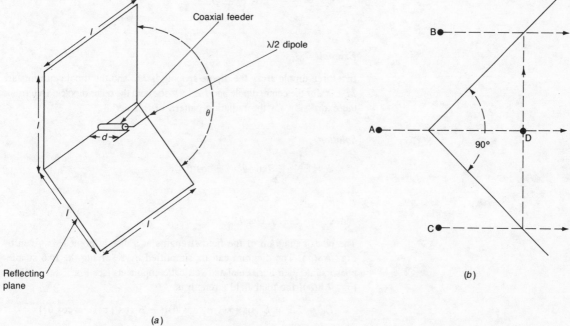

Fig. 7.7 (*a*) Corner reflector, and (*b*) operation of the corner reflector.

either solid metal or wire mesh or perhaps a system of conductors spaced apart at 0.1λ or less, and they are at an angle of either $90°$ or $60°$ to one another. Each plane must be at least one wavelength long in each direction, i.e. $l \geq \lambda$. The spacing of the $\lambda/2$ dipole from the intersection of the planes is somewhere between 0.3λ and 0.5λ with the latter distance being the most common. Assuming that the plane angle θ is $90°$, Fig. 7.7(*b*) shows that reflections from the corner reflector produce image dipoles A, B and C; image A has the same polarity as the actual dipole D, while both images B and C have the opposite polarity. Effectively, therefore, the corner reflector gives a four-dipole array with two dipoles in each of two planes. The way in which the array factor of such an array can be calculated is shown on page 146.

Three-dipole Array

An increase in both the gain and the directivity can be obtained if a third dipole is added to give a three-dipole array. If the centre dipole B is taken as the reference then at the distant point P the field strength due to dipole A will lag the field strength due to aerial B by an angle

$$\psi = \frac{2\pi d}{\lambda} \cos \theta \pm \alpha;$$

while the field strength due to dipole C will lead by an angle

$$\psi^1 = \frac{2\pi d}{\lambda} \cos \theta \pm \beta.$$

Example 7.3

In a three-dipole array the dipole spacing is $\lambda/2$ and the dipole currents are $2I \angle 0°$ for the centre dipole and $I \angle 180°$ for both the outer dipoles. Determine the expression for the radiation pattern of this aerial.

Solution

$$\phi = \frac{2\pi}{\lambda} \frac{\lambda}{2} \cos \theta = \pi \cos \theta.$$

$$\psi = \pi - \pi \cos$$

and $\psi^1 = \pi + \pi \cos \theta.$

The phasor diagram of the field strengths at a distant point P is given by Fig. 7.8(*a*). The diagram can be simplified by resolving the two smaller phasors into their horizontal and vertical components (see Fig. 7.8(*b*)). From Fig. 7.8(*b*), the total field strength is

$$E_T = 2E + E \cos [\pi(1 + \cos \theta)] + E \cos [\pi(1 - \cos \theta)]$$

$$= 2E[1 - \cos (\pi \cos \theta)] = 4E \sin^2 [(\pi/2) \cos \theta]$$

or $\quad E_T = \dfrac{4}{\sqrt{3}} E_0 \sin^2 [(\pi/2) \cos \theta].$ (*Ans.*)

Broadside Array

A broadside array consists of a number n of $\lambda/2$ dipoles equally spaced in one line and carrying equal-amplitude, in-phase currents. If the total power supplied to the array is P_t watts, each dipole will be fed

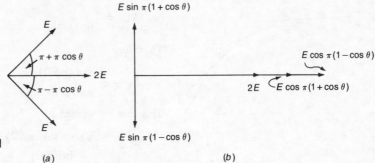

Fig. 7.8 (*a*) Phasor diagram of a three-dipole array, and (*b*) simplified version of (*a*).

(*a*)   (*b*)

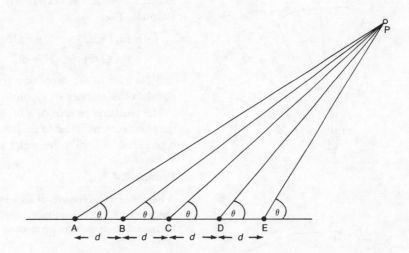

Fig. 7.9 Broadside array.

with P_t/n watts and will produce a field strength of $E_0/\sqrt{n}$ at the distant point P.

Suppose, for example, that the number n of dipoles is 5, as in Fig. 7.9. The radiation from dipole A has a further distance $d \cos \theta$ to travel in order to reach the distant point P than has the radiation from dipole B. The field strength due to aerial A will therefore lag the field strength due to dipole B by angle $\phi = (2\pi d/\lambda) \cos \theta$ radians. In similar manner, field strength E_B lags field strength E_C, E_C lags E_D and E_D lags E_E, all by the same angle ϕ. Figure 7.10(*a*) shows the phasor diagram of the field strengths at point P.

If a line is drawn normal to the centre of each phasor the lines will meet at the point marked as 0. Lines then drawn from point 0 to the ends of each phasor subtend the angle ϕ (see Fig. 7.10(*b*)). The total field strength E_T is the phasor going from the base of phasor E_E to the tip of phasor E_A as shown by Fig. 7.10(*c*). From this diagram, $E_T = 2AC \sin (n\phi/2)$. To find the value of the length AC, consider the triangle ACD. From this

$$\sin (\phi/2) = \frac{E/2}{AC}$$

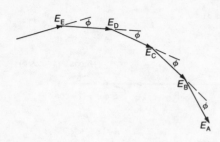

(a)

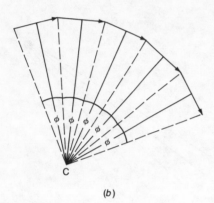

(b)

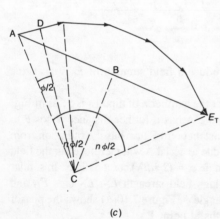

(c)

Fig. 7.10 Phasor diagrams of the field strengths produced by a broadside array.

or $AC = \dfrac{E/2}{\sin(\phi/2)}$.

Therefore

$$E_T = \frac{E \sin(n\phi/2)}{\sin(\phi/2)}. \tag{7.5}$$

This can be written as

$$E_T = \frac{(E_0/\sqrt{n}) \sin(n\phi/2)}{\sin(\phi/2)} = \frac{E_0\sqrt{n} \sin(n\phi/2)}{n \sin(\phi/2)}. \tag{7.6}$$

When the angle θ is very nearly equal to $90°$, $\cos\theta$, and hence ϕ, is small. Then

$$\frac{\sqrt{n} \sin(n\phi/2)}{n \sin(\phi/2)} \simeq \frac{\sqrt{n}\, n\phi/2}{n\, \phi/2} = \sqrt{n}$$

and the gain of the broadside array is $(\sqrt{n})^2$ or n, i.e. the gain is equal to the number of dipoles in the array.

The radiation pattern of a broadside array has two main lobes in the plane perpendicular to the line of the array. Its exact shape is best determined by finding the angles at which maxima and minima occur.

Example 7.4

A broadside array consists of six vertical $\lambda/2$ dipoles spaced $\lambda/2$ apart, and energized by equal-amplitude, in-phase currents. Calculate and plot the horizontal-plane radiation pattern of the aerial.

Solution
Here $n = 6$ and

$$\phi = \frac{2\pi}{\lambda}\frac{\lambda}{2}\cos\theta = \pi\cos\theta,$$

and so

$$E_T = \frac{E_0}{\sqrt{6}}\frac{\sin(3\pi\cos\theta)}{\sin[(\pi/2)\cos\theta]}.$$

At $\theta \simeq 90°$

$$E_T = \frac{E_0}{\sqrt{6}} \times 6 = \sqrt{6}E_0.$$

Nulls occur in the radiation pattern when the numerator of equation (7.6) is zero and the denominator is *not* zero, i.e. when $3\pi\cos\theta = \pi, 2\pi, 3\pi$, etc. Hence

(a) $\cos\theta = \frac{1}{3}$, $\theta = \pm70.5°$

(b) $\cos\theta = \frac{2}{3}$, $\theta = \pm48.2°$

(c) $\cos\theta = 1$, $\theta = 0°$.

The centre of each minor lobe occurs when $\sin(n\phi/2) = 1$ or $n\phi/2 = \pm(2k + 1)\pi/2$ or

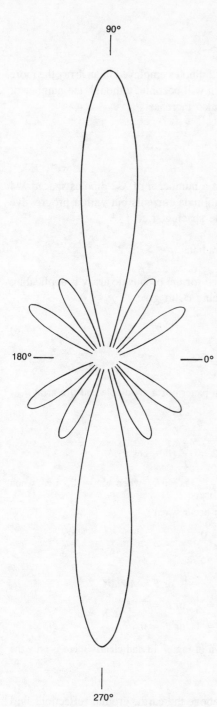

$$\cos \theta = \frac{\pm(2k + 1)\lambda}{2nd}.$$

In this case,

$$\cos \theta = \frac{\pm(2k + 1)\lambda}{2 \times 6 \times \lambda/2} = \frac{\pm(2k + 1)}{6}:$$

(a) $k = 1$; $\theta = \cos^{-1}[\pm\frac{1}{2}] = \pm60°$;

(b) $k = 2$; $\theta = \cos^{-1}[\pm\frac{5}{6}] = \pm33.6°$;

(c) $k = 3$; $\theta = \cos^{-1}[\pm\frac{7}{6}]$; which is, of course, not possible.

The radiation pattern is shown by Fig. 7.11.

Beamwidth

The beamwidth of an aerial is the angle subtended by the 3 dB points on the major lobe(s) of the radiation pattern. Thus, in Fig. 7.12 the angle β is the beamwidth. The maximum field strength produced by a broadside array is $E_{T(max)} = \sqrt{n}E_0 = nE$. Therefore

$$\frac{E_T}{E_{T(max)}} = \frac{\sin(n\phi/2)}{n\sin(\phi/2)}.$$

If the angle θ that gives $E_T = E_{T(max)}/\sqrt{2}$ is large, $\cos\theta$ will be small and then $\sin(\phi/2)$ is small. Therefore

$$\frac{E_T}{E_{T(max)}} \simeq \frac{\sin(n\phi/2)}{n\,\phi/2}$$

and this is equal to $1/\sqrt{2}$ when $n\phi/2 = 1.39$ radians. Hence

$$\frac{n\pi d \cos\theta}{\lambda} = 1.39$$

or $\cos\theta = \sin(90° - \theta) = \dfrac{1.39}{n\pi d}$.

Since θ is large, $\sin(90° - \theta) \simeq \theta = \beta/2$, and so the beamwidth β is

$$\beta = \frac{0.885\lambda}{nd} = \frac{51\lambda^0}{nd}. \tag{7.7}$$

In the case of the array in Example 7.4

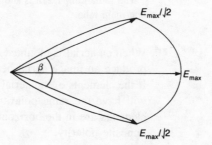

Fig. 7.11 Broadside-array radiation pattern.

Fig. 7.12 Beamwidth of an aerial.

$$\beta = \frac{51 \times \lambda}{6 \times \lambda/2} = 17°.$$

The greater the number of $\lambda/2$ dipoles employed in an array the more directive the radiation pattern will become, although the number of small secondary lobes will also increase.

End-fire Array

An end-fire array consists of a number n of $\lambda/2$ dipoles spaced $\lambda/4$ apart and fed with equal-amplitude currents but with a progressive phase difference equal to the spacing, i.e.

$$\alpha = \frac{2\pi}{\lambda} \lambda/4 = \pi/2 \text{ radians.}$$

The same analysis as that used for the broadside array is applicable, but with $\psi = \phi + \alpha$ replacing ϕ, to give

$$E_T = \frac{\sqrt{n}E_0 \sin (n\psi/2)}{n \sin (\psi/2)}. \tag{7.8}$$

Example 7.5

Calculate and plot the radiation pattern of a six-dipole end-fire array.

Solution

$$\psi = \frac{2\pi}{\lambda} \frac{\lambda}{4} \cos \theta + \pi/2 = \frac{\pi}{2} (1 + \cos \theta).$$

The maximum field strength $E_{T(max)}$ is $\sqrt{6}E_0$ when $\psi = 0$, i.e. when $(\pi/2)(1 + \cos \theta) = 0$ or $\theta = 180°$.

Nulls in the radiation pattern occur when

$$\sin [(3\pi/2)(1 + \cos \theta)] = 0$$

or $\quad (3\pi/2)(1 + \cos \theta) = k\pi.$

(a) $k = 1$; $\cos \theta = \frac{2}{3} - 1 = -\frac{1}{3}$ or $\theta = 109.5°$.

(b) $k = 2$; $\cos \theta = \frac{4}{3} - 1 = \frac{1}{3}$ or $\theta = 70.5°$.

(c) $k = 3$; $\cos \theta = \frac{6}{3} - 1 = 1$ or $\theta = 0°$.

The radiation pattern is shown in Fig. 7.13 and clearly there is only the one main lobe.

Height Factor

When an aerial is mounted above the earth, ground reflections will produce an image aerial beneath the earth as shown by Fig. 7.14. If the elements of the aerial are mounted vertically the image aerial will have the same polarity as the actual aerial, but if the aerial elements are in the horizontal plane the image aerial will be of the opposite polarity.

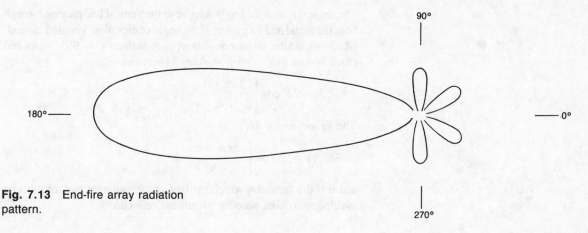

Fig. 7.13 End-fire array radiation pattern.

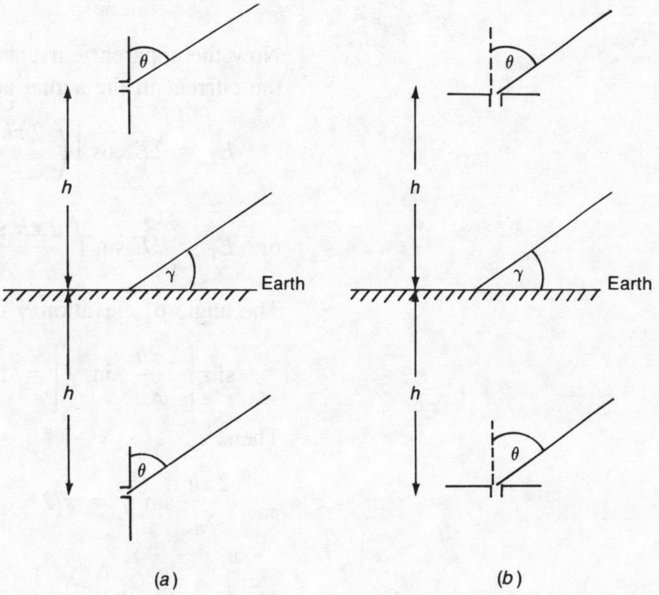

Fig. 7.14 Aerial mounted above the earth: (*a*) vertical aerial, and (*b*) horizontal aerial.

(*a*)　　　　　(*b*)

An aerial and its image behave in the same way as a two-dipole array with a spacing equal to $2h$, where h is the height of the aerial above the earth.

Vertical Aerial

The aerial and image currents are in phase with one another, and so the total field strength produced at a distant point P in the vertical plane is (from equation (7.1) with $d = 2h$)

$$E_T = 2E \cos\left(\frac{2\pi h}{\lambda} \cos\theta\right).$$

The angle θ is now the angle relative to the vertical line passing through both the aerial and its image. If the angle of elevation γ is used instead, which would be more convenient in practice, $\gamma = 90° - \theta$, and $\cos \theta = \cos (90° - \gamma) = \sin \gamma$. Therefore

$$E_T = 2E \cos \left[\frac{2\pi h \sin \gamma}{\lambda} \right]. \tag{7.9}$$

The *height factor* $H(\gamma)$ is

$$H(\gamma) = 2 \cos \left[\frac{2\pi h \sin \gamma}{\lambda} \right]$$

and it is the factor by which the free-space radiation pattern must be multiplied to take account of ground reflections.

Horizontal Aerial

Now the apparent current in the image aerial is in anti-phase with the current in the actual aerial, and hence

$$E_T = 2E \cos \left[\left(\frac{2\pi h \sin \gamma}{\lambda} \right) + 90° \right]$$

or $\quad E_T = 2E \sin \left(\frac{2\pi h \sin \gamma}{\lambda} \right). \tag{7.10}$

The angle of elevation γ of the main beam is the one that makes

$$\sin \left[\frac{2\pi h}{\lambda} \sin \gamma \right] = 1.$$

Then

$$\frac{2\pi h}{\lambda} \sin \gamma = \pi/2$$

or $\quad \gamma = \sin^{-1} \left[\frac{\lambda}{4h} \right].$

Example 7.6

A transmitting aerial operates at 20 MHz and it is to have its maximum radiation in the vertical plane at an angle of elevation of 14°. Determine the height above ground at which the aerial should be mounted if it is (*a*) horizontally, or (*b*) vertically mounted.

Solution

$$\lambda = \frac{3 \times 10^8}{20 \times 10^6} = 15 \text{ m.}$$

(a) From equation (7.10), the maximum field strength is obtained when

$$\sin\left[\left(\frac{2\pi h}{\lambda}\right)\sin 14°\right] = 1$$

or $\dfrac{2\pi h}{\lambda}\sin 14° = \pi/2$.

Therefore,

$$h = \frac{15}{4\sin 14°} = 15.5 \text{ m.} \quad (Ans.)$$

(b) From equation (7.9), the maximum field strength occurs at the height where

$$\cos\left[\left(\frac{2\pi h}{\lambda}\right)\sin 14°\right] = -1$$

or $\left(\dfrac{2\pi h}{\lambda}\right)\sin 14° = \pi$.

Therefore

$$h = \frac{15}{2\sin 14°} = 31 \text{ m.} \quad (Ans.)$$

Figure 7.15 shows the height factor $H(\gamma)$ plotted for heights of $\lambda/4$, $\lambda/2$, $3\lambda/4$ and λ above the earth for both horizontal and vertical aerials.

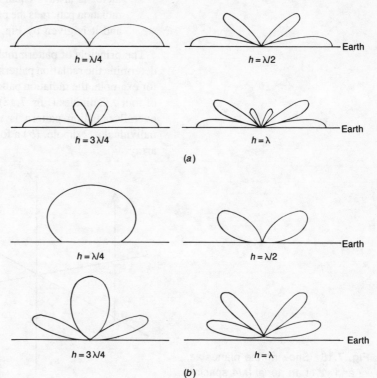

Fig. 7.15 Height factor for (a) vertical and (b) horizontal aerials.

It is clear that with a horizontal aerial there is zero radiation at ground level as well as at some angles of elevation.

Pattern Multiplication

The use of *pattern multiplication* to obtain the radiation pattern of a two-dipole array has already been touched upon (p. 136), but it was not specifically labelled as such. The radiation pattern of a two-dipole array can be drawn for each of the three planes shown in Fig. 7.16.

(a) In the plane xz, i.e. the equatorial plane, the array factor is $\sqrt{2}\cos[(\pi/4)\cos\theta]$. When $\theta = 0°$ this gives $\sqrt{2}\cos(\pi/4) = 1$, and when $\theta = 90°$ it gives $\sqrt{2}\cos 0° = \sqrt{2}$ (see Fig. 7.17(a)(i)). The radiation pattern of a single dipole is a circle (Fig. 7.17(a)(ii)), and the overall radiation pattern in this plane is the product of (a)(i) and (a)(ii) and this is shown by Fig. 7.17(a)(iii).

(b) In the plane xy, the meridian plane, the array factor is the same as for the equatorial plane. The dipole pattern is given by equation (6.19) and it is shown by Fig. 7.17(b)(ii). The overall radiation pattern is given by the product of these figures and it is shown by Fig. 7.17(b)(iii).

(c) In the plane yz, once again each dipole has a radiation pattern given by Fig. 7.17((b)(ii)=(c)(ii)) but in the array factor equation the angle θ has only the value of 90°. Hence, the array factor is always equal to $\sqrt{2}$ (see Fig. 7.17(c)(i)). The overall radiation pattern is the product of Figs 7.17(c)(i) and 7.17(c)(ii) and it is given by Fig. 7.17(c)(iii).

The principle of pattern multiplication can always be employed to determine the radiation patterns of complex aerial systems. Suppose, for example, the radiation pattern of an array consisting of three rows of four $\lambda/2$ dipoles (Fig. 7.18) is to be determined. In each plane the overall radiation pattern is the product of (a) the pattern of an individual $\lambda/2$ dipole, (b) a four-dipole array, and (c) a three-dipole array.

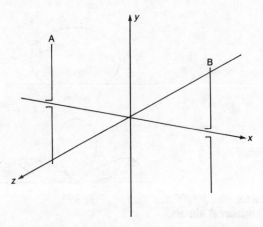

Fig. 7.16 Showing the planes xz, xy and yz of an aerial ($\lambda/4$ spacing).

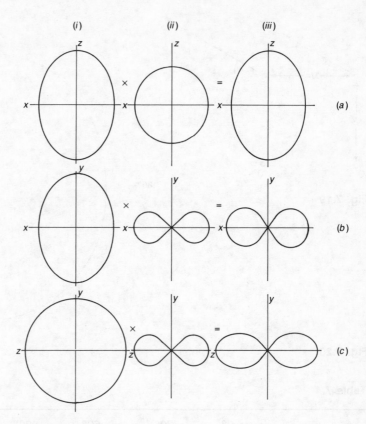

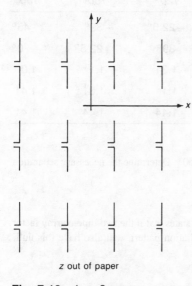

Fig. 7.17 Pattern multiplication.

Fig. 7.18 4 × 3 array.

Example 7.7

Calculate and plot the horizontal-plane radiation pattern of the four vertical $\lambda/2$ dipole array shown in Fig. 7.19. The dipoles are fed with equal-amplitude, in-phase currents.

Solution

The radiation pattern due to dipoles A and B, and to C and D, is given by

$$E_T = \sqrt{2} E_0 \cos \left[\frac{2\pi\lambda}{4 \times 2\lambda} \cos \theta \right] = \sqrt{2} E_0 \cos \left[\frac{\pi}{4} \cos \theta \right].$$

Similarly, the radiation pattern due to the dipoles A and C, or B and D, is given by

$$E_T = \sqrt{2} E_0 \cos \left[(\pi/4) \sin \theta \right].$$

The overall radiation pattern is therefore given by

$$E_T = 2 E_0^2 \cos \left[(\pi/4) \cos \theta \right] \cos \left[(\pi/4) \sin \theta \right].$$

This is tabulated in Table 7.4. The radiation pattern is shown in Fig. 7.20.

Example 7.8

A dipole array is receiving a strong interference signal 50° from the centre of the main lobe of its radiation pattern. To eliminate this interference another

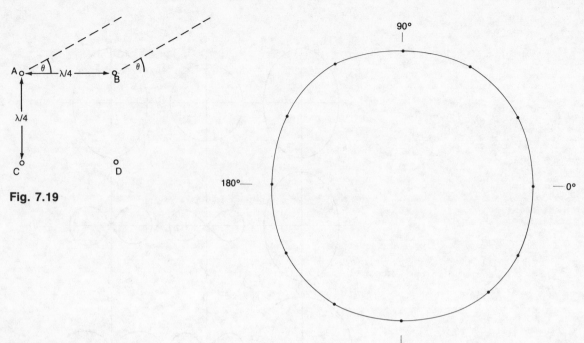

Fig. 7.19

Fig. 7.20

Table 7.4

θ	0°	30°	60°	90°	120°	150°	180°
45° cos θ	45°	39°	22.5°	0°	−22.5°	−39°	−45°
45° sin θ	0°	22.5°	39°	45°	39°	22.5°	0°
$\sqrt{2}$ cos (45° cos θ)	1.0	1.1	1.31	1.41	1.31	1.1	1.0
$\sqrt{2}$ cos (45° sin θ)	1.41	1.31	1.1	1.0	1.1	1.31	1.41
Overall	1.41	1.44	1.44	1.41	1.44	1.44	1.41

array is to be used to position a null at 50°. Determine the necessary separation of the two arrays.

Solution
The principle of pattern multiplication states that if the two-dipole array factor has a null at 50° then the overall radiation pattern will also have this null. Hence

$$\cos \left[\frac{\pi d}{\lambda} \cos 50° \right] = 0$$

or $$\frac{0.643 \, \pi d}{\lambda} = \pi/2$$

and $$d = \frac{\lambda}{2 \times 0.643} = \frac{\lambda}{1.286} = 0.78 \, \lambda. \quad (Ans.)$$

Mutual Impedances between Dipoles

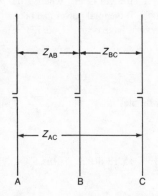

Fig. 7.21 Mutual impedances between dipoles.

In the treatment so far of dipole arrays it has been assumed that there was zero interaction between the dipoles. This assumption is not correct if the spacing between the dipoles is small. Each dipole that carries a current will induce a voltage into every other nearby dipole and this means that each pair of dipoles has a mutual impedance between them. The magnitude and phase of each mutual impedance is determined by the length and the diameter of the two dipoles and by their spacing.

When a current flows in a dipole the e.m.f. it induces into another dipole in the array tends to oppose the voltage that is applied to that dipole. The voltage applied to a dipole must therefore be equal to the sum of (*a*) the voltage needed to produce the dipole current if there was zero mutual impedance, and (*b*) the voltages necessary to balance the e.m.f.s induced into the dipole by currents flowing in the other dipoles. Consider, as an example, the three-dipole array shown in Fig. 7.21. For this array the voltage equations are:

$$
\begin{aligned}
V_A &= I_A Z_{AA} + I_B Z_{AB} + I_C Z_{AC} \\
V_B &= I_B Z_{BB} + I_A Z_{AB} + I_C Z_{BC} \\
V_C &= I_C Z_{CC} + I_A Z_{AC} + I_B Z_{BC}.
\end{aligned}
\tag{7.11}
$$

The input impedance of a dipole is the ratio (input voltage)/(input current); hence

$$
\begin{aligned}
Z_A &= V_A/I_A = Z_{AA} + Z_{AB} I_B/I_A + Z_{AC} I_C/I_A \\
Z_B &= V_B/I_B = Z_{BB} + Z_{AB} I_A/I_B + Z_{BC} I_C/I_B \\
Z_C &= V_C/I_C = Z_{CC} + Z_{AC} I_A/I_C + Z_{BC} I_B/I_C.
\end{aligned}
\tag{7.12}
$$

The self-impedance, Z_{AA}, Z_{BB}, Z_{CC}, of a dipole is equal to its radiation resistance, which if the dipole is slightly shorter than $\lambda/2$, is 73.14 Ω.

If the dipoles are fed with equal-amplitude, in-phase currents, as for a broadside array, then $I_A = I_B = I_C$ and $Z_A = Z_{AA} + Z_{AB} + Z_{AC}$. If the dipoles are spaced $\lambda/4$ apart and the currents have a progressive phase difference of 90°, as for an end-fire array, then $I_B = jI_A$ and $I_C = jI_B = j^2 I_A = -I_A$, and, for example, $Z_A = Z_{AA} + jZ_{AB} - Z_{AC}$.

The power radiated by each dipole is equal to the square of the dipole current times the real part of the dipole impedance. Thus, the power P_A radiated by dipole A is equal to $I_A^2 \times$ (real part of Z_A).

Example 7.9

Two vertical $\lambda/2$ dipoles spaced $\lambda/4$ apart are supplied with equal-amplitude currents having a phase difference of 90°. Each dipole has a radiation resistance of 73 Ω and their mutual impedance is $40 - j30\ \Omega$. Calculate (*a*) the radiated power, and (*b*) the gain of the array in dBi, if the dipole currents are both 100 mA.

Solution

(*a*) $I_A = 100$ mA, $I_B = j100$ mA.

$$Z_A = 73 + (40 - j30) \times \frac{j100}{100} = 103 + j40 \ \Omega.$$

$$P_A = (100 \times 10^{-3})^2 \times 103 = 1.03 \ W.$$

$$Z_B = 73 + (40 - j30) \times \frac{-j100}{100} = 43 - j40 \ \Omega.$$

$$P_B = (100 \times 10^{-3})^2 \times 43 = 0.43 \ W.$$

The total radiated power = 1.46 W.   (*Ans.*)

(*b*) In the end-fire direction the electric field strength is $2E$, where E is the field strength produced by either dipole. If the total power had been supplied to one dipole only its current would have been $\sqrt{(1.46/73)} = $ 141.4 mA, giving a field strength of

$$\frac{141.4}{100} \times E.$$

Therefore, the gain, relative to a $\lambda/2$ dipole, is

$$20 \log_{10} \left[\frac{2E \times 100}{141.4E} \right] = 3.02 \ dB.$$

Therefore, gain in dBi = 3.02 + 2.16 = 5.18 dBi.   (*Ans.*)

Example 7.10

Determine the gain, relative to a $\lambda/2$ dipole, of a four-dipole broadside array that uses a dipole spacing of $\lambda/4$. The mutual impedances between the dipoles are $Z_{AB} = Z_{BC} = Z_{CD} = -12 + j30 \ \Omega$, $Z_{AC} = Z_{BD} = 4 + j20 \ \Omega$, and $Z_{AD} = -2 - j12 \ \Omega$. Each dipole has a radiation resistance of 73 Ω.

Solution

The aerial currents are of equal amplitude and phase. Hence,

$$Z_A = 73 + (-12 + j30) + (4 + j20) + (-2 - j12).$$

The real part of Z_A is

$$73 - 12 + 4 - 2 = 63 \ \Omega$$

and so the radiated power is $P_A = 63I^2$ W. Also,

$$Z_B = 73 + (-12 + j30) + (-12 + j30) + (4 + j20),$$

giving a real part of 53 Ω. The power radiated by dipole B is $P_B = 53I^2$ W.

$$Z_C = 73 + (4 + j20) + (-12 + j30) + (-12 + j30)$$

and the radiated power is $P_C = 53I^2$ W. Lastly,

$$Z_D = Z_A = 63 + j38 \ \Omega$$

and the radiated power $P_D = 63I^2$ W.

The total radiated power is

$$P_T = P_A + P_B + P_C + P_D = 232I^2 \ W.$$

If this power were radiated by one dipole only, its current would have to be $\sqrt{(232I^2/73)} = 1.783I$. Therefore

$$\text{the gain} = 20 \log_{10}\left[\frac{4E}{1.783E}\right] = 7.02 \text{ dB.} \quad (Ans.)$$

If the mutual impedances had been ignored the gain of the array would have been $20 \log_{10} \sqrt{4} = 6.02$ dB.

It should be noted that when the dipoles are supplied with in-phase currents the power radiated by each dipole can be obtained by taking only the resistive part of each mutual impedance.

The Rhombic Aerial

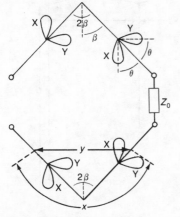

Fig. 7.22 Rhombic aerial.

A long-wire radiator (p. 128) produces a radiation pattern that has one main lobe and a number of small side-lobes. The main lobe is at an angle θ to the wire axis and is a function of the electrical length of the wire. The rhombic aerial uses four long-wire radiators to form, in the horizontal plane, a rhombus shape as shown by Fig. 7.22. The *tilt angle β* is chosen to ensure that (*a*) the lobes marked as X point in opposite directions and their radiated energies cancel out, and (*b*) the lobes marked as Y point in the same direction and hence their radiations are additive. For this to occur two things are necessary: (*a*) the distance x from the mid-point of one wire to the mid-point of the next must be $\lambda/2$ longer than the direct distance y between these two points; (*b*) the tilt angle β should be equal to $(90° - \theta)$. Since the lobe angle varies with frequency the choice of the tilt angle must be a compromise and it is generally calculated at the geometric mean of the two extreme operating frequencies.

The gain of a rhombic aerial is a function of the length of each wire, the tilt angle and the angle of elevation. The higher the gain that is required the smaller must be the angle of elevation but, fortunately, this is a favourable situation for long-distance routes. Typically, a rhombic aerial will operate over a 2:1 frequency ratio, e.g. 7–14 MHz, with a gain that varies from about 3 dB to about 15 dB. Since the h.f. band covers a frequency ratio of about 4:1, three, or four, rhombic aerials would be needed to give complete coverage.

Although the rhombic aerial was very widely employed in the past it has now been superseded for most applications by the log-periodic aerial. The log-periodic h.f. aerial can provide a wider bandwidth than can the rhombic aerial and it is physically smaller.

The Log-periodic Aerial

The log-periodic aerial (l.p.a.) is a type of aerial whose radiation pattern and gain change very little over a wide frequency band. The bandwidth of the aerial is restricted only by the physical size of the elements at the low-frequency end of the aerial, and by the accuracy of the construction at the upper-frequency end. Several different forms of l.p.a. exist but perhaps the most common consists of a tapered dipole array. Figure 7.23 shows the basic form of a dipole l.p.a.; moving along the aerial from the feed point both the length of each

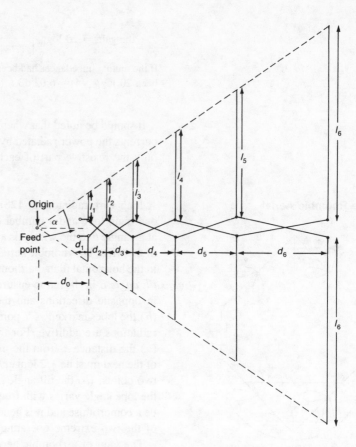

Fig. 7.23   Basic log-periodic aerial.

dipole, and the spacing between adjacent dipoles, increase by a constant ratio. This ratio, τ, is known as the *scale factor*.

$$\tau = \frac{l_2}{l_1} = \frac{l_3}{l_2} = \frac{l_n}{l_{n-1}}. \tag{7.13}$$

This means, of course, that $l_2 = l_1\tau$, $l_3 = l_2\tau = l_1\tau^2$, and so on up to $l_n = l_1/(\tau^{n-1})$. The aerial is fed, not at the origin, but at a point distance $d_1 = d_2/\tau$ to the left of the first dipole. The distance d_0 from the origin to the first dipole is then equal to $d_1/(1 - 1/\tau)$. The *characteristic angle* α is, from Fig. 7.24, obtained from

$$\tan \alpha = \frac{l_1}{d_0} = \frac{l_2}{d_0 + d_1}, \text{ etc.}$$

The *space factor* σ is the name given to $\tan \alpha$.

At any frequency within the operating bandwidth of the aerial only three, or perhaps four, of the dipoles are at, or near, the resonant length of $\lambda/2$. Only these dipoles will radiate appreciable power and they are said to be in the *active region*. The active region forms a radiation centre whose dimensions, in wavelengths, are both constant and independent of frequency. All other dipoles, which are either much longer or much shorter than $\lambda/2$ will radiate little, if any, energy.

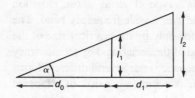

Fig. 7.24

As the frequency of the current supplied to the l.p.a. is varied the position of the active region will move back and forth along the array. The result is that both the gain and the radiation pattern of the l.p.a. remain approximately constant with frequency.

The highest, and the lowest, operating frequencies coincide with those wavelengths which make the shortest, and the longest, dipoles $\lambda/2$ in length. The difference in length may easily be as much as 10:1 and the ratio gives the corresponding aerial bandwidth. Thus the maximum frequency is $f_{max} = c/\lambda_{min} = c/4l_1$, the minimum frequency is $f_{min} = c/\lambda_{max} = c/4l_n = c/(l_1/4\tau^{n-1})$. Hence

$$\frac{f_{max}}{f_{min}} = \frac{1}{\tau^{n-1}}.$$

The practical bandwidth obtained is somewhat less than this ratio would predict because otherwise the active region would sometimes move right off the array.

The excitation of the dipoles in the l.p.a. is always achieved by feeding the current in at the smaller-dimensioned end, and transposing the dipoles along the line of the array. This is necessary to ensure an end-fire radiation pattern in the direction of the origin. Figure 7.25 shows typical radiation patterns for both the horizontal and the vertical planes at two different frequencies. The radiation pattern has a wide main lobe and it is clearly not very directive (less so than a rhombic aerial), but it does have small side-lobes.

A horizontally polarized l.p.a. may be used in various configurations, either supported from a tower or suspended from masts. Figure 7.26 shows two practical high-frequency log-periodic aerials; the vertically polarized version in (a) has a very low angle of elevation and this makes it suitable only for long-distance radio links. The aerial gain is reduced because the ground in the vicinity of the aerial is not of zero resistance. Because of their larger gain most h.f. l.p.a.s are horizontally polarized. The aerial cannot be mounted at a fixed physical height above the ground because its electrical height would

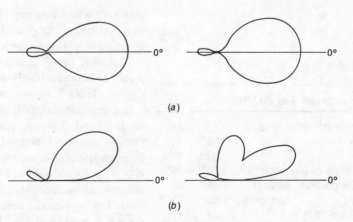

Fig. 7.25 Log-periodic aerial radiation pattern: (a) horizontal plane, and (b) vertical plane.

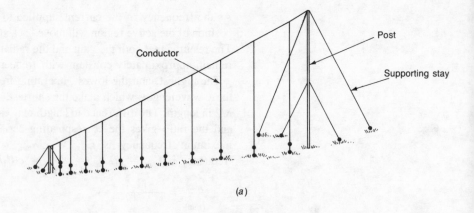

(a)

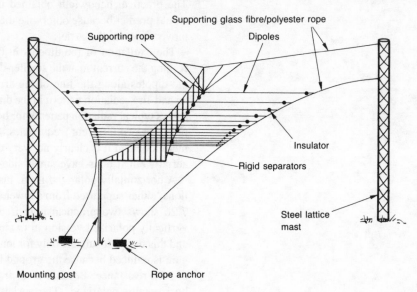

(b)

Fig. 7.26 (a) Vertically polarized l.p.a., (b) Horizontally polarized l.p.a.

Table 7.5

Bandwidth 4 to 28 MHz		
Number of dipoles	17	46
α	18°	10°
τ	0.85	0.95
Angle of elevation	25°	25°
Beamwidth, vertical	45°	25°
horizontal	60°	150°
Gain (dBi)	12	15

then vary with frequency. This is why horizontally polarized l.p.a.s slope relative to the ground to ensure that each dipole is at the same electrical height above earth. Figure 7.26(b) shows a typical example.

Typically, an h.f. l.p.a. can operate over most of the h.f. band with a gain of between 10 dBi and 15 dBi, depending on the number of dipoles. Table 7.5 gives two sets of typical data for l.p.a.s.

Log-periodic aerials are widely employed for various services in the h.f. band. They are particularly useful when a wide horizontal plane beamwidth is wanted, when a high elevation angle is needed, or when site area is a problem. An increased gain can be obtained if two l.p.a.s are used in a broadside array. If a number of areas at different azimuth angles are to be served a rotatable l.p.a. is often used. The log-periodic aerial is also often used in both the v.h.f. and the u.h.f. bands and Fig. 7.27 shows a typical example.

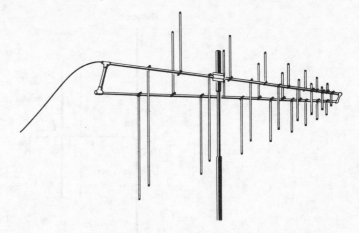

Fig. 7.27 v.h.f./u.h.f. log-periodic aerial.

The Yagi Aerial

The equatorial and meridian plane radiation patterns of a $\lambda/2$ dipole, shown in Figs 6.12(a) and (b), are not directive enough for many applications. An increase in both the gain and the directivity can be obtained by the addition of one, or more, *parasitic elements*. A parasitic element is one that is not directly supplied with an exciting current and that is coupled by mutual impedance to the driven $\lambda/2$ dipole. If the parasitic element is longer than $\lambda/2$ and it is mounted behind the dipole, relative to the required direction of maximum radiation, it is known as a *reflector*. Conversely, if the parasitic element is shorter than $\lambda/2$ and it is mounted in front of the dipole it is known as a *director*. A *Yagi aerial* consists of a $\lambda/2$ dipole, a reflector, and one, or more, directors. Figures 7.28(a) and (b) show, respectively, a $\lambda/2$ dipole with one reflector, or with one director. A parasitic element can be directly fixed onto a common metal support since its mid-point is a voltage node.

The spacing between the $\lambda/2$ dipole and each of the parasitic elements is small enough for there to be mutual impedances between them. When a current is supplied to the dipole and causes it to radiate energy an e.m.f. will be induced into each parasitic element. This e.m.f. will make a current flow in the element so that the parasitic element accepts power from the dipole and then re-radiates it. The phase relationships between the energy radiated by the $\lambda/2$ dipole and the energy radiated by each parasitic element depends upon both the element spacing and the phase of the current in each element. In turn, the phase of the current in an element is determined by the electrical length of that element. The element spacings and lengths are chosen to give the maximum radiation in the wanted direction and minimum radiation in all other directions.

A director increases the radiated field on its side of the dipole while a reflector concentrates the radiated field in the opposite side of the dipole. Since a reflector is longer than $\lambda/2$ in length it has an inductive reactance, while, conversely, a director has a capacitive reactance. Values of reactance for different lengths of element are shown by

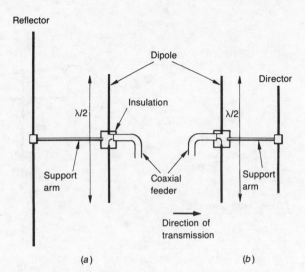

Fig. 7.28 (*a*) Dipole with reflector.
(*b*) Dipole with one director.

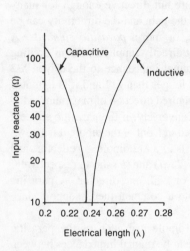

Fig. 7.29 Reactance/electrical-length curves for a dipole.

Fig. 7.29; note that when the length is $\lambda/2$ the reactance of the dipole is $+j43 \ \Omega$.

Consider a $\lambda/2$ dipole and a reflector. Zero voltage is applied to the reflector and hence equation (7.11) becomes

$$V_D = I_D Z_{DD} + I_R Z_{DR} \tag{7.14}$$

$$0 = I_D Z_{DR} + I_R Z_{RR}. \tag{7.15}$$

From equation (7.15), $I_R = -I_D Z_{DR}/Z_{RR}$, and hence

$$V_D = I_D \left[Z_{DD} - \frac{Z_{DR}^2}{Z_{RR}} \right]$$

or $Z_D = Z_{DD} - \dfrac{Z_{DR}^2}{Z_{RR}}.$ $\tag{7.16}$

If typical values for Z_{DD}, Z_{RR} and Z_{DR} are put into equation (7.16) it will be found that the input impedance of the dipole has been considerably reduced. This reduction, which is accentuated if one, or more, directors are added, may lead to difficulties in matching the aerial to the coaxial feeder. To overcome this difficulty a *folded dipole* is often used (Fig. 7.30). This has an impedance of about 300 Ω.

Example 7.11

An aerial consists of a $\lambda/2$ dipole of $73 + j0 \ \Omega$ impedance and a parasitic element whose impedance is $73 + j68 \ \Omega$. If the mutual impedance between the dipole and the parasitic element is $65 + j0 \ \Omega$, calculate (*a*) the input impedance and (*b*) the front-to-back ratio of the aerial. The dipole−reflector spacing is 0.13λ.

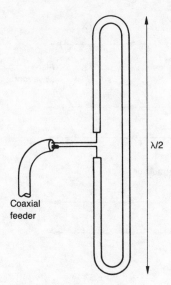

Fig. 7.30 Folded dipole.

Solution

(*a*) From equation (7.16)

$$Z_D = 73 - \frac{65^2}{73 + j68} = 42 + j29 \ \Omega. \quad (Ans.)$$

(*b*) $I_R = \dfrac{-I_D Z_{DR}}{Z_{RR}},$

and therefore

$$\frac{I_R}{I_D} = \frac{-65}{73 + j68} = 0.65 \angle 137°.$$

In the direction from reflector to dipole there is a phase difference of $137° - 360° \times 0.13 = 90°$ and hence the field strength is proportional to $\sqrt{(1^2 + 0.65^2)} = 1.194$. In the opposite direction, i.e. from dipole to reflector, the phase difference is $137° + 360° \times 0.13 = 180°$. Hence, the field strength in this direction is proportional to $1 - 0.65 = 0.35$. Therefore

$$\text{the front-to-back ratio} = 20 \log_{10} \left[\frac{1.194}{0.35} \right] = 10.66 \ \text{dB}. \quad (Ans.)$$

The addition of more directors will increase the gain of the Yagi aerial, and Fig. 7.31 shows the relationship between the number of directors and the gain. The element spacing is most critical for the dipole and the reflector, and for the dipole and the first director; the former should normally be somewhere in the range 0.17λ to 0.2λ.

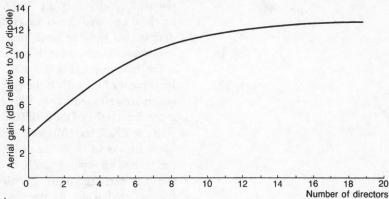

Fig. 7.31 Showing the relationship between the gain of a Yagi array and the number of directors.

The director spacings are usually chosen to be in the range 0.15λ to 0.4λ, however a large number of other combinations of lengths/spacings are equally likely to be used.

The Parabolic Dish Aerial

For point-to-point radio links in the upper part of the u.h.f. band and in the s.h.f. band the usual aerial employed is the *parabolic dish*. The

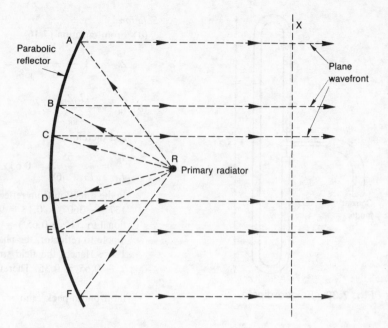

Fig. 7.32 Principle of a parabolic dish aerial.

parabolic dish aerial is essentially a large metal dish that is illuminated by a source of spherical wavefront, and which converts the incident radiation into a wave having a plane wavefront. It is necessary to control the amplitude of the dish illumination from its centre to its edge both to maximize the aerial gain and to minimize losses. The *illumination efficiency* η is 100 times the ratio of the on-axis directivity for a given illumination to the directivity produced by a uniform illumination with the same total radiated power. This means that a uniformly illuminated dish has an illumination efficiency of 100%.

The basic concept of a dish aerial is illustrated by Fig. 7.32. The focal point of the dish is the point at which incident parallel waves which are reflected by the dish converge. The feed point is placed at the focal point of the dish; energy is radiated from the feed point with a spherical wavefront and is directed on to the dish. The geometry of the dish is such that the distance from the focal point to the dish and then to an arbitrary plane X on the other side of the focal point is a constant, *regardless* of which point on the surface of the dish is considered. Thus, the distance RAX = distance RBX = RCX = RDX = RFX. This ensures that the spherical wavefront produced by the feed is converted into a plane wavefront at the plane X. The reflected waves are all parallel with one another and form a highly directive radio wave.

The effective aperture A_e of the dish is, for an illumination efficiency of 100%, equal to the geometric aperture, i.e.

$$A_e = \pi \left(\frac{D}{2}\right)^2,$$

where D is the diameter of the dish. The gain G of the aerial, with respect to an isotropic radiator is, from equation (6.24),

$$G = \frac{\pi (D/2)^2}{\lambda^2/4\pi} = \frac{\pi^2 D^2}{\lambda^2} \tag{7.17}$$

or, relative to a $\lambda/2$ dipole

$$G = \frac{\pi (D/2)^2}{0.13\lambda^2} = 6\left(\frac{D}{\lambda}\right)^2. \tag{7.18}$$

Clearly, for a dish aerial to have a high gain its diameter must be several times greater than the signal wavelength. This is the reason why the dish aerial is only employed in the u.h.f. and s.h.f. bands.

The radiation pattern of a dish aerial is highly directive with a very narrow main lobe in the direction of the dish axis. Consequently, the usual way of drawing the radiation pattern gives insufficient detail. Normally, therefore, the radiation pattern is drawn using Cartesian co-ordinates as shown by Fig. 7.33.

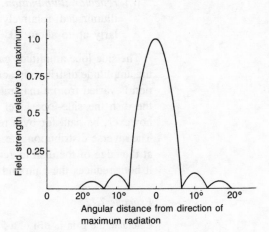

Fig. 7.33 Radiation pattern of a parabolic dish aerial.

The beamwidth of a dish aerial is approximately given by equation (7.19), i.e.

$$\text{beamwidth} = \frac{70\lambda^\circ}{D}. \tag{7.19}$$

Side Lobes

The radiated power contained in the side lobes of the aerial's radiation pattern is power that is radiated in unwanted directions where it may well interfere with other systems. The net efficiency of a dish aerial can be increased if the total energy in the side lobes is minimized. For the ground aerial of a communications satellite system

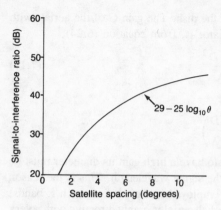

Fig. 7.34 Side-lobe specification for a ground-station aerial.

the side-lobe performance, in the direction of the geometric arc, is specified by such bodies as CCIR, INTELSAT and EUTELSAT. Ninety per cent, or more, of the side lobes are required to lie under the envelope defined by $29 - 25 \log_{10} \theta$, where θ is the side-lobe angle in degrees. This relationship gives the signal-to-noise ratio in decibels for various satellite spacings shown by the graph of Fig. 7.34. The design of a ground-station aerial must reduce the side-lobe level to the defined figure.

There are a number of sources of side-lobe energy produced by a dish aerial. These are:

(a) *feed spillover* due to direct radiation from the feed, and/or (for a Cassegrain or Gregorian aerial) direct radiation from the sub-reflector, spilling past the edge of the main reflector;

(b) *blockage* in which the feed point, or the sub-reflector, blocks some of the energy reflected by the main reflector, thus producing a hole in the energy distribution of the aerial's aperture; and

(c) *reflector illumination*, where, if the main dish is uniformly illuminated, relatively large side lobes are produced, particularly up to about 10° from the aerial's axis.

The side-lobe amplitude can be considerably reduced by tapering the amplitude distribution across the dish. If the amplitude distribution is varied from a maximum at the centre, to zero at the edge of the dish the side-lobe level will be reduced. This reduction will, however, be paid for by a reduction in the illumination efficiency. The inverse distribution, i.e. varying illumination from a maximum at the edge of the dish to zero at the centre, is not employed since it both reduces the gain and increases the side-lobe level.

Example 7.12

Calculate the gain in dBi of a 5 m diameter dish aerial at (a) 4 GHz, and (b) 10 GHz if the illumination efficiency is 60%. (c) The gain of an INTELSAT standard dish aerial is 60 dBi; calculate its diameter at 11 GHz if the illumination efficiency is 60%.

Solution

(a) $\lambda = \dfrac{3 \times 10^8}{4 \times 10^9} = 0.075$ m.

The gain $= 10 \log_{10} \left[0.6 \left(\dfrac{5\pi}{0.075} \right)^2 \right] = 44.2$ dBi. (*Ans.*)

(b) $\lambda = \dfrac{3 \times 10^8}{10 \times 10^9} = 0.03$ m.

Gain $= 10 \log_{10} \left[0.6 \left(\dfrac{5\pi}{0.03} \right)^2 \right] = 52.2$ dBi. (*Ans.*)

(c) $\lambda = 0.027$ m.

$$60 \text{ dBi} = 10^6 = 0.6\left[\frac{\pi D}{0.027}\right]^2$$

$$\text{or } D = \sqrt{123} \simeq 11.1 \text{ m.} \quad (Ans.)$$

Feed Arrangements

The simplest form of primary feed is a dipole fed by a coaxial line and, usually, with a parasitic element, or a plane reflector, to ensure that the radiation is directed onto the dish. At frequencies greater than a few gigahertz a waveguide feeder will be employed to connect the aerial to the transmitter (or receiver). Then, it is more usual to employ some kind of waveguide feed and one commonly employed example is the horn radiator. The horn radiator is just a length of waveguide whose open end is flared out to form a horn.

There are four main ways in which the main reflecting dish can be illuminated by the feed: these are known, respectively, as *front-feed* (or *focus*), *Cassegrain*, *Gregorian* and *offset-feed*. The basic arrangement of each of these methods is shown in Fig. 7.35.

Front feed

Front feeding a dish aerial, shown by Fig. 7.35(a), is the simplest method of illuminating the main reflector. To reduce the side-lobes to an acceptable level the illumination efficiency is only about 55 to 60%. The feed and its supporting structure produces aperture blockage and this increases the side-lobe level. There may also be some re-radiation from the struts, which will degrade the *cross-polarization*† of the aerial. A horn front feed can give a good gain and an acceptable noise temperature and it is often used with small aerials.

Cassegrain

The Cassegrain method of feeding a dish aerial, shown in Fig. 7.35(b), uses a sub-reflector to reflect the energy radiated by the primary feed onto the main reflector. The system has the advantages over the front-feed method in that the feed system is simpler and a shorter length of waveguide feeder is necessary. Although aperture blockage and strut re-radiation are still a problem the Cassegrain aerial is commonly employed.

---

†Communications satellite systems often employ identical frequency bands to carry different signals. Two signals are transmitted at the same frequency but one is horizontally polarized and the other is vertically polarized. The cross-polarization isolation is a measure of the degree of crosstalk immunity between the two signals.

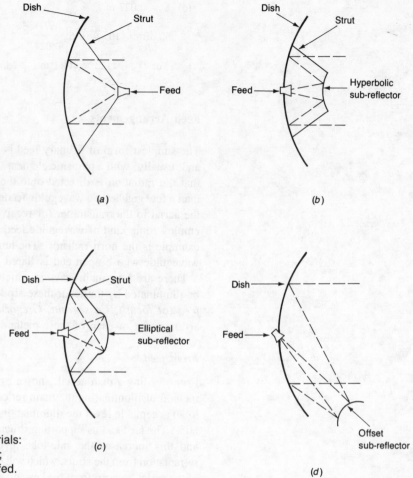

Fig. 7.35 Parabolic dish aerials:
(*a*) front-feed; (*b*) Cassegrain;
(*c*) Gregorian; and (*d*) offset-fed.

Gregorian

The Gregorian aerial, Fig. 7.35(*c*), is a variant of the Cassegrain aerial in which an elliptical sub-reflector is employed. This gives an increased illumination efficiency of about 76% and also improved cross-polarization isolation.

Offset Feed

Blockage of the transmitted beam may be avoided if the feed, or the sub-reflector, is mounted outside the area occupied by the main beam. Figure 7.35(*d*) shows an offset Cassegrain dish aerial. A front-feed, or a Gregorian aerial can be similarly offset. The feed, and the sub-reflector, are mounted at the appropriate angles to the aerial axis to ensure that the main beam is directed along the axis. The overall size of the offset aerial is smaller, for a given gain at a given frequency, and the side lobes are smaller, but de-polarization effects are increased. This type of aerial is becoming increasingly popular.

8 Propagation of Radio Waves

The radio-frequency energy radiated by a transmitting aerial in the direction(s) determined by its radiation pattern will travel through the atmosphere and arrive at the distant aerial by one, or more, different modes of propagation. These are the *ground wave*, the *sky wave*, the *space wave*, *tropospheric scatter*, and via a *communications satellite*. When the transmit aerial is mounted on the surface of the earth propagation will mainly be by means of the ground or surface wave. There will also be some sky-wave radiation which may, or may not, be returned to earth. The ground wave is used in the v.l.f. and l.f. bands for long-distance, narrow-bandwidth communications, and in the m.f. band for shorter-distance sound-broadcast signals. In the v.l.f. band the received signal at long distances from the transmitter is extremely stable and shows very little change over a day, or even seasonally. The main disadvantage of the v.l.f. band is its very limited bandwidth and the large physical structures that are necessary for the transmitting aerials. The range of the ground wave is not quite as large for signals in the l.f. and the m.f. bands since the attenuation suffered increases with frequency; the daytime range is typically a few hundred kilometres at the low end of the m.f. band to about 100 km at the upper end of the m.f. band. The night-time range may well be some thousands of kilometres because propagation can then take place via the ionosphere.

At higher frequencies the transmit aerial can be mounted above the earth, perhaps by several wavelengths, and then propagation is mainly by either the sky wave or the space wave. The sky wave is mainly employed for long-distance telephony links in the h.f. band but also for some sound-broadcast services. The space wave is employed for various services in the v.h.f., u.h.f. and s.h.f. bands, such as point-to-point multi-channel telephony systems and land-, sea- and air-mobile systems. The u.h.f. band is also utilized for television-broadcast signals.

The scope of tropospheric scatter systems is much more limited and it finds application for some beyond-the-horizon radio systems where considerable terrain difficulties exist. Finally, communications satellite systems are now employed to carry wideband telephony and television systems over long distances.

The Ionosphere

Ultraviolet, and other, radiation from the sun ionizes large numbers of the atoms which make up gas molecules in the upper atmosphere.

This ionization causes the earth to be surrounded by a belt of ionized gases which is known as the *ionosphere*. The height of the ionosphere above the surface of the earth varies considerably but it is normally within the limits of 50 km to 400 km. Within the ionosphere the density of the free electrons is much higher than it is at heights either above or below the ionosphere. Below the ionosphere there are two other regions: the *troposphere*, from ground level to about 10 km above the ground; and the *stratosphere*, from about 10 km to the lower edge of the ionosphere.

The electron density, measured by the number of free electrons per cubic metre, within the ionosphere is not a constant quantity. During the daytime the ionosphere contains four distinct regions, or layers, within which the electron density exhibits a maximum value. These four layers are known as the D, E, F_1 and F_2 layers, in order of their increasing height above ground, and their increasing electron density. There is also a fifth layer, labelled C, below the D layer, but it plays no part in radio-communications. The heights of the four layers above the ground vary daily, seasonally, and with the 11-year variations in the 22-year sun-spot cycle. Approximate heights, for both daytime and night-time, are shown by Figs 8.1(a) and (b), respectively. In the daytime the D layer has a much lower electron density than do the other layers, and during the night it disappears completely. The E layer becomes weaker at night but it does not normally disappear. In the daytime the F_1 layer is at a height of about 200 km but the height of the F_2 layer varies considerably. Typically, the F_2 layer is at about 250 km to 350 km in the winter, and between 300 km and 400 km in the summer.

Refractive Index of the Ionosphere

As a radio wave propagates through the ionosphere its electric field will exert a force upon the free electrons which causes them to move. If the free electron density is N electrons per m^3, each free electron has a charge of e coulombs, and the field strength of the radio wave

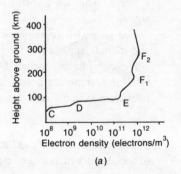

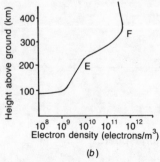

Fig. 8.1 Ionospheric layers (a) in the daytime, and (b) at night.

(a)

(b)

is $E \sin \omega t$ radians, then the force exerted upon each free electron is $F = eE \sin \omega t$ newtons. Each electron is given an acceleration of

$$\frac{F}{m} = \frac{eE \sin \omega t}{m},$$

where m is the mass of an electron.

The consequent velocity v of an electron is obtained by integrating its acceleration with respect to time; hence

$$v = (-eE \cos \omega t)/\omega m \text{ m/s.}$$

The current represented by this movement of electrons is

$$i = Ne \times (-eE \cos \omega t)/\omega m = (-Ne^2 E \cos \omega t)/\omega m \text{ A/m}^2.$$

The current that flows in a dielectric of unity relative permittivity is

$$i = \epsilon_0 \, dE/dt = \epsilon_0 \omega E \cos \omega t \text{ A/m}^2$$

and so the total current flow is

$$i = \epsilon_0 \omega E \cos \omega t - \frac{Ne^2 E \cos \omega t}{\omega m}$$

$$= \epsilon_0 \left[1 - \frac{Ne^2}{\omega^2 m \epsilon_0} \right] \omega E \cos \omega t.$$

From this, the relative permittivity ϵ_r of an ionospheric layer is

$$\epsilon_r = 1 - \frac{Ne^2}{\omega^2 m \epsilon_0}.$$

Substituting in the values for the charge and the mass of an electron gives

$$\epsilon_r = 1 - \frac{81N}{f^2}.$$

This means that the relative permittivity of an ionospheric layer is smaller than that of free space.

The refractive index n of a layer is equal to the square root of its relative permittivity, and hence

$$\text{Refractive index } n = \sqrt{\left(1 - \frac{81N}{f^2} \right)}. \tag{8.1}$$

Also, of course, the refractive index is equal to the ratio (sine of angle of incidence)/(sine of angle of refraction).

Behaviour of the Ionosphere at Different Frequencies

The effect of the ionosphere on an incident radio wave is very much a function of frequency. In the v.l.f. and l.f. bands the ionosphere has a high conductivity and it *reflects*, with very little loss, any wave that is incident upon its lower edge. In the m.f. band the D layer acts like a lossy medium whose attenuation reaches its maximum value at the *gyro-frequency* of 1.4 MHz. Signals in the m.f. band are absorbed by the D layer and are not returned to earth. At frequencies in the h.f. band the E and F layers *refract* radio waves and, if conditions are correct, return the wave to earth. The D layer has little, if any, refractive effect. Signals in the v.h.f., u.h.f. and s.h.f. bands normally pass straight through the ionosphere.

Ionospheric Variations

The intensity of the ultraviolet, and other, radiation from the sun that enters the earth's atmosphere is continually fluctuating. Both regular and irregular variations occur. The regular variations occur for two reasons: (*a*) the intensity of the sun's radiation varies with both the time of the day and the month in the year; and (*b*) sun spots occur at the sun's surface which produce 11-year fluctuations in its radiation that affect the ionosphere. The sun-spot cycle has an average periodicity of twenty-two years.

Irregular ionospheric disturbances are also experienced. Solar flares emit large amounts of radiation from the sun that produce a large increase in the ionization of the D layer. This may cause the D layer to absorb all h.f. signals, giving a complete blackout for anything up to about two hours. Sometimes, ionospheric storms occur; this is the name given to irregular fluctuations in the conductivity of the ionosphere which cause rapid fading, particularly at the higher frequencies. Ionospheric storms tend to occur at intervals of 27 days.

Sporadic E consists of a cloud of drifting electrons which suddenly, and unpredictably, appear within the E layer. The electron cloud has a much higher electron density than is usual and it is therefore able to return to the earth waves that normally pass straight through the ionosphere. When sporadic E is present, the m.u.f. of a returned wave may be, typically, some 20−40 MHz, although even higher figures sometimes occur. Sporadic E is more likely to occur in the summer than in the winter.

The Troposphere

The CCIR have adopted the following expression for the refractive index n of the troposphere.

$$n = 1 + \frac{77.6}{T}\left(p + \frac{4810e}{T}\right) \times 10^{-6},\tag{8.2}$$

where T is the temperature in K, and e and p are the water-vapour and atmospheric pressures in millibars. Since all three parameters are functions of height this expression can be written in the form

$$n(h) = 1 + ae^{-bh},\tag{8.3}$$

where a and b are constants, e is the base of natural logarithms, i.e. e = 2.7183, and h is the height above the ground in kilometres. If $a = 315 \times 10^{-6}$ and $b = 0.136$ (which refer to the CCIR *average atmosphere*), the CCIR reference refractive index is obtained. The refractive index, calculated from equation (8.3) gives most inconvenient numbers, e.g. at $h = 1$ km, $n = 1.000\ 27$, and at $h = 2$ km, $n = 1.000\ 24$. To obtain more convenient numbers it is customary to use the *refractivity* of the troposphere instead. This is merely equal to $(n - 1) \times 10^{6}$; thus for $h = 1$ km the refractivity is equal to 270. The variation of the refractivity with increase in height above the ground is shown by Fig. 8.2.

Since the refractive index falls with increase in height the radio wave will follow a curved path through the troposphere. Provided the wave is launched in a horizontal plane the radius of curvature of the path is $-\mathrm{d}h/\mathrm{d}n$. In the British Isles, at a height of about 1 km, n is normally about 40 parts in 10^{6} lower than at ground level (see Fig. 8.2); hence $-\mathrm{d}h/\mathrm{d}n = 2.5 \times 10^{7}$ m. The radius of the earth is 6400 km and so the radius of curvature of the radio path is 25000/6400 = 3.9 times the radius of the earth. When a radio link is planned it is convenient to consider that the radius of the earth is k times its actual value and that the space wave travels in a straight line (see Fig. 8.3). Under normal atmospheric conditions $k = 4/3$, giving an effective earth's radius of $(4 \times 6400)/3 \simeq 8500$ km.

If the distance between the two aerials is not too great the earth between them may be assumed to be flat. Little error is then introduced into any calculations and, if required, a correction factor can be used. Departures from normal atmospheric conditions are not usually large enough to give significant fading on a line-of-sight radio path. Sometimes, however, abnormal conditions do arise that noticeably affect space-wave propagation.

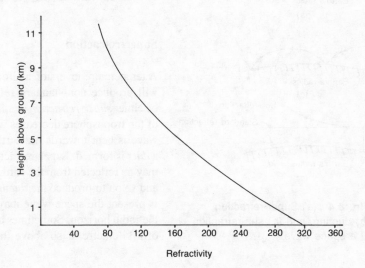

Fig. 8.2 Variation of refractivity with height.

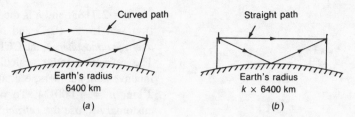

Fig. 8.3 Space-wave propagation.

Temperature Inversion

In the early part of a sunny day the cloudless skies may result in the air temperature being higher than the ground temperature, contrary to the normal state of affairs. This is known as a *temperature inversion*. Temperature inversions also occur because of the following.

(a) *Subsidence:* a mass of warm air may be further heated by compression and then rises to a greater height, while cooler air falls to a lower height to replace it.

(b) *Dynamic:* a mass of warm air may move on top of a mass of cold air.

(c) *Nocturnal:* the air at the surface of the earth is rapidly cooled after sunset.

(d) *Cloud layer:* the sun's rays may be reflected from the upper surface of a cloud and heat up the air above the cloud.

Water Vapour

For normal atmospheric conditions the humidity of the atmosphere falls gradually with increase in height, but sometimes an abrupt change in the humidity gradient may occur. This is most likely to happen above the sea during hot weather and it often occurs to the leeward of land as warm air moves from the land out over the sea.

Super-refraction

A temperature inversion and/or a non-standard water-vapour gradient will produce non-standard refraction of the space wave. This may be either *super-refraction* or *sub-refraction*. When the refractive index of the troposphere decreases with height more rapidly than usual the wave is bent towards the earth to a greater extent than normal and a *duct* is formed. Super-refraction is shown by Fig. 8.4(a). The wave may be reflected from the earth, again super-refracted, again reflected and so on to produce the *ducting* shown in Fig. 8.4(b). When a duct is present the space wave may propagate for distances well beyond the radio horizon. Sometimes an elevated duct may appear at a height of a kilometre or so above the ground.

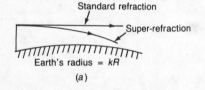

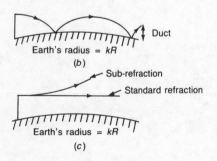

Fig. 8.4 (a) Super-refraction, (b) ducting, and (c) sub-refraction of a space wave.

Sub-refraction

Sub-refraction of a space wave occurs when the refractive index of the troposphere decreases with height at a smaller rate than is usual. The space wave is then refracted to a lesser extent than normal, and if the k factor of 4/3 has been employed in the design of the link, the wave will appear to be bent upwards as shown by Fig. 8.4(c).

Ground-wave Propagation

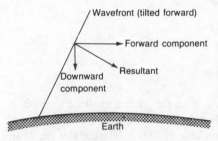

Fig. 8.5 Ground-wave propagation.

At very low, and low frequencies the transmitting aerial is an electrically short monopole which radiates energy in the form of a ground, or surface, wave. The ground wave is vertically polarized and it is able to follow the undulations of the surface of the earth because of diffraction.† As the wave travels, its magnetic field cuts the earth and induces e.m.f.s into it. In turn, these induced e.m.f.s cause currents to flow in the resistance of the earth and so dissipate power. This power can only be supplied by the radio wave and so there is a continuous flow of energy from the wave into the ground. This results in the wavefront having two components of velocity, one in the forward direction and one downwards normal to the earth. This is shown by Fig. 8.5; the resultant velocity of the wave is the phasor sum of the two components and this makes the wavefront tilt forwards. Since the downward component of velocity is always normal to the surface of the earth the wave is able to follow the undulations of the ground.

The wave is attenuated, for two reasons, as it travels. First, the wavefront diverges as it travels so that the field strength is inversely proportional to distance and, second, power is taken from the wave to supply the ground losses. The calculation of the ground power losses is complex and it depends upon such factors as the frequency of the wave, and the conductivity and permittivity of the earth. The attenuation is expressed by an attenuation factor K whose value can be approximately predicted from published graphs. Thus, the electric field strength E_D at a distance D kilometres from the transmitting aerial is given by

$$E_D = \frac{KE_1}{D} \text{ V/m}, \tag{8.4}$$

where E_1 is the electric field strength 1 km from the transmitting aerial. E_1 is equal to $300\sqrt{P_t}$ (p. 118), where P_t is the transmitted power in kilowatts. Hence, if the distance D is expressed in kilometres

$$E_D = \frac{K\,300\sqrt{P_t}}{D} \text{ mV/m}. \tag{8.5}$$

---

†Diffraction is a phenomenon which occurs with all wave motion. It causes a radio wave to bend around any obstacle it passes. For the ground wave the earth itself is the obstacle.

At low, and medium frequencies the attenuation factor K is inversely proportional to the square of the frequency. Hence the attenuation of the wave increases rapidly with increase in frequency. At high frequencies and above the (frequency) factor reduces the amplitude of the ground wave to a negligible value even over quite small distances.

Example 8.1

A radio transmitter is to produce, using the ground wave, a field strength of 5 mV/m at a distance of 200 km. If the attenuation factor is 0.2 calculate the necessary radiated power. What would be the field strength if the frequency of the transmitted signal were to be doubled?

Solution

$$E_{200} = 5 = \frac{0.2 \times 300\sqrt{P_t}}{200}$$

or $P_t = 278$ kW.   (*Ans.*)

If the frequency is doubled

$$E_{200} = 1.25 \text{ mV/m.}$$ (*Ans.*)

At frequencies in the m.f. band the maximum field strength at ground level is obtained when the height of the transmitting aerial is $5\lambda/8$. There is then appreciable radiation into the sky (p. 119). During the daytime, the sky wave is completely absorbed by the D layer and does not return to the earth. During the night, however, the D layer has disappeared and then the sky wave will be returned to earth via either the E or the F layer. The two cases of interest are illustrated by Figs 8.6(*a*) and (*b*). In the first case, a transmitted signal is received at a distant location by means of both the ground wave and the sky wave. The total field strength at this point is the phasor sum of the individual field strengths and this will vary because of fluctuations in the length of the ionospheric path. This means that the received signal will be prone to fading. In the second case, two different signals, radiated at the same frequency by different transmitters, are received together at a point. The quality of the received wanted signal will be impaired and its reception may be so poor that

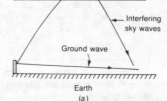
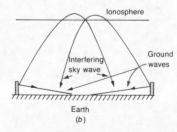

Fig. 8.6 Medium-frequency transmissions at night.

it is unusable. This is, of course, the effect that makes the night-time reception of medium-wave broadcast signals in Europe of such poor quality.

Sky-wave Propagation

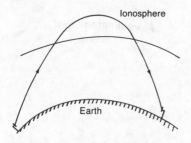

Fig. 8.7 Sky-wave propagation.

The basic principle of a radio link operating in the h.f. band is illustrated by Fig. 8.7. The sky wave is directed into the ionosphere where it is continuously refracted. If, before it reaches the top of the F_2 layer, it has been refracted to the extent that the angle of refraction is 90° then the wave will be returned to earth. The intrinsic instability of the ionosphere causes the length of a sky-wave path to vary continuously in a random manner, and considerable fading may take place. Particularly difficult in this respect are the periods around dawn and around dusk, when the electron densities of the ionosphere change more rapidly than at other times.

The potential unreliability of an h.f. radio link has meant that h.f. radio has, in the past, lost considerable ground to both communications satellite and terrestrial radio-relay systems. Nowadays, however, the relative cheapness of h.f. radio systems plus the introduction of various technical innovations, such as diversity, frequency synthesis, and new modulation techniques, have revived interest in h.f. technology. Congestion in the h.f. band has been partially alleviated by an increased use of s.s.b./i.s.b. transmissions.

A radio wave that enters the E layer with an angle of incidence ϕ_i will continuously be refracted away from the normal. If the values of the electron density N and the frequency f are such that $\sin \phi_i = \sqrt{(1 - 81N/f^2)}$ then, since the refractive index $n = (\sin \phi_i)/(\sin \phi_r)$, $\sin \phi_r$ must be equal to unity. Then $\phi_r = 90°$ and the wave must then be travelling in a horizontal direction. Any further refraction of the wave will then return it back to earth. If no part of the E layer has an electron density large enough for the $\sin \phi_r = 1$ relationship to be satisfied, the sky wave will not be returned to earth but will escape from the top of the layer. The wave will then be incident on the F_1 (or, at night, the F) layer with an increased angle of incidence and it will here be further refracted and may perhaps be returned to earth by this layer. If not, the wave will leave the top of the F_1 layer and pass on to the F_2 layer with an even larger angle of incidence and now it may be returned to earth. If the wave is not returned by the F_2 layer it will escape from the earth. The concept is illustrated by Figs 8.8(a) and (b). In Fig. 8.8(a) a wave entering the E layer with an angle of incidence ϕ_1 is returned to earth but another signal, at the same frequency, which is incident on the E layer with a smaller angle of incidence ϕ_2, is not returned. This second wave travels on to the F_1 layer and is from here returned to earth. Figure 8.8(b) shows two waves of frequencies f_1 and f_2 incident upon the E layer with the same angle of incidence ϕ_1, where $f_2 > f_1$. The lower-frequency wave is returned to earth by the F_1 layer but the higher-frequency wave is not.

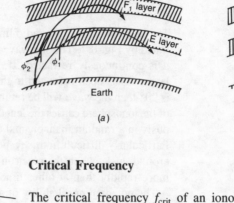

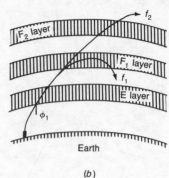

Fig. 8.8 Showing the effect on a sky wave of (a) the angle of incidence, and (b) frequency.

(a)

(b)

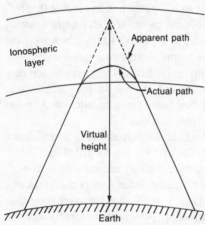

Fig. 8.9 Virtual height of a layer.

Critical Frequency

The critical frequency f_{crit} of an ionospheric layer is the highest frequency that can be radiated upwards with an angle of incidence ϕ_i of zero and be returned to earth. This will be the frequency of the wave that travels up to the top of the layer, where the electron density is at its highest value, before it is refracted to the extent that ϕ_r becomes equal to 90°. From equation (8.1),

$$\sin 0° = 0 = \sqrt{(1 - 81 N_{max}/f_{crit}^2)}$$

or

$$f_{crit} = 9\sqrt{N_{max}}. \tag{8.6}$$

Each of the ionospheric layers will have its own value of critical frequency. From the ground it appears as though the wave has travelled in a straight line, has been reflected by the ionosphere, and has then returned to earth from the point of reflection in another straight-line path. The *virtual height* of a layer is the height at which this apparent reflection takes place (see Fig. 8.9).

Maximum Usable Frequency

The maximum usable frequency (m.u.f.) of a layer is the highest frequency that can be employed for a sky-wave path between two points on the earth's surface. For a wave to be returned to earth $\sin \phi_r = 1$, and so

$$\sin \phi_i = \sqrt{(1 - f_{crit}^2/f_{max}^2)};$$

$$1 - \sin^2 \phi_i = \frac{f_{crit}^2}{f_{max}^2}$$

or

$$f_{max} = \text{m.u.f.} = \frac{f_{crit}}{\cos \phi_i} = f_{crit} \sec \phi_i. \tag{8.7}$$

Since the electron density of each layer is subject to continuous fluctuations, some regular and predictable and others not, the m.u.f.

is not a constant figure. The m.u.f. of a sky-wave path between two points on the earth's surface will vary throughout the day, and graphs of the forecast m.u.f. for various routes are commercially available.

Optimum Working Frequency

Because of the instability of the ionosphere, operating a radio link at a frequency equal to the m.u.f. would not give a reliable system. It is customary to work an h.f. route at a frequency of about 80% of the m.u.f.; this lower frequency is known either as the *optimum working frequency* or as the *optimum traffic frequency*.

Since the m.u.f. is not of constant value it will be necessary to be able to change the frequency of a sky-wave link as, and when, the propagation conditions alter. Usually, an h.f. radio transmitter is allocated (although not exclusively) several different frequencies, and any one of them may be in use at a given time. When the propagation conditions are severe it may become necessary to use two, or more, frequencies simultaneously and, perhaps, even retransmit when conditions improve.

Example 8.2

The virtual height of a layer is 110 km and its critical frequency is 4 MHz. Calculate the m.u.f. for two points on the surface of the earth that are 600 km apart if (*a*) the earth is assumed to be flat, and (*b*) the radius of the earth is 6400 km.

Solution

(*a*) From Fig. 8.10(*a*),

$$\phi = \tan^{-1}\left(\frac{300}{110}\right) = 70°.$$

Therefore

the m.u.f. $= 4 \sec 70° = 11.7$ MHz.   (*Ans.*)

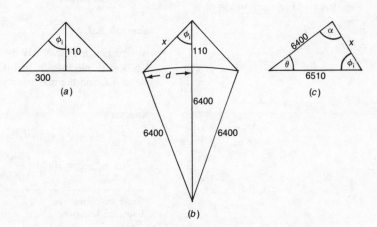

Fig. 8.10

(b) From Fig. 8.10(b) and then Fig. 8.10(c), $d = 300 = 6400\theta$. $\theta = 300/6400 = 0.0469$ radians $= 2.69°$.

$$x^2 = 6400^2 + 6510^2 - (2 \times 6400 \times 6510 \times \cos 2.69°),†$$

or $x = 322$ km. Therefore,

$$\frac{6400}{\sin \phi_i} = \frac{x}{\sin 2.69°},‡$$

$$\sin \phi_i = \frac{6400 \sin 2.69°}{322}$$

and $\phi_i = 68.9°$, and

the m.u.f. $= 4 \sec 68.9° = 11.1$ MHz.   (*Ans.*)

Maximum Value of ϕ_i

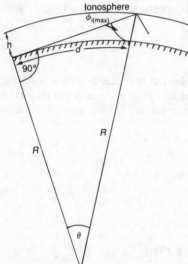

Ionosphere

$\phi_{i(max)}$

h

d

90°

R

R

θ

Fig. 8.11 Maximum value of ϕ_i.

The angle of incidence ϕ_i with which a sky wave enters the ionosphere cannot be increased without limit. The maximum possible value $\phi_{i(max)}$ occurs when the transmitted wave is tangential to the earth's surface, as shown by Fig. 8.11. Here R is the radius of the earth, approximately 6400 km, and h is the virtual height of a layer. From the figure,

$$\phi_{i(max)} = \sin^{-1}\left(\frac{R}{R + h}\right). \tag{8.8}$$

Also, $\theta = 180° - (90° + \phi_{i(max)}) = 90° - \sin^{-1}[R/(R + h)] = \cos^{-1}[R/(R + h)]$. Therefore, the maximum ground range is

$$2d = 2R\theta = 2R\cos^{-1}\left(\frac{R}{R + h}\right). \tag{8.9}$$

The maximum distance using the E layer is about 4000 km; longer distances are possible using the F layer(s). If a path length approaching, or exceeding, the maximum range is wanted a multi-hop path will be necessary.

Example 8.3

At a particular time of day the E layer has a maximum electron density of 1.5×10^{11} electrons/m and it is at a virtual height of 150 km. Calculate the m.u.f. and the maximum single-hop range.

Solution

$$f_{crit} = 9\sqrt{(1.5 \times 10^{11})} = 3.49 \text{ MHz}.$$

$$\sin \phi_{i(max)} = \sin^{-1}\left(\frac{6400}{6550}\right) = 77.7°.$$

---

†Using the cosine rule.
‡Using the sine rule.

Therefore, the m.u.f. is 3.49 sec 77.7° = 16.4 MHz.   (*Ans.*)

The maximum ground range $= 2 \times 6400 \times \cos^{-1}\left(\dfrac{6400}{6550}\right)$

$$= 2745 \text{ km.} \quad (Ans.)$$

Skip Distance

There is a maximum usable frequency for any distance and this distance is also the minimum distance at which that frequency can be transmitted over the sky-wave path. This minimum distance is called the *skip distance*.

Fading

General fading, in which the complete signal fades to the same extent, is produced by fluctuations in the ionospheric attenuation. Unless there is a complete fade-out of the signal the effects of general fading can be overcome by the use of a.g.c. in the radio receiver.

Selective fading occurs when the signal picked up by the receive aerial has arrived via two, or more, different paths (see Fig. 8.12). The total field strength at the aerial is the phasor sum of the field strengths produced by each signal. The phase difference between the signals arriving via the two separate paths is equal to $2\pi/\lambda$ times the difference between the lengths of the two paths. If this difference should vary, due to fluctuations in the ionosphere, the total field strength will also vary in a frequency-dependent manner because of the $1/\lambda$ term. This means that the different frequency components of a complex signal may fade to different extents.

There are a number of ways in which selective fading may be combatted. These include:

(*a*) the use of a highly directive transmitting aerial so that the number of possible propagation paths is minimized;

(*b*) operation at a frequency as near to the m.u.f. as possible;

(*c*) the use of s.s.b./i.s.b. signals;

(*d*) Lincompex; and

(*e*) the use of space and/or frequency diversity.

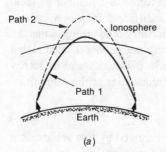

Path 2 ⟍   Ionosphere

Path 1

Earth

(*a*)

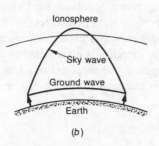

Ionosphere

Sky wave

Ground wave

Earth

(*b*)

Fig. 8.12 Selective fading.

Example 8.4

The signals received by an aerial arrive over two different paths, one of which is 75 km longer than the other. At the carrier frequency the two signals cancel out. If the carrier is amplitude modulated, at what side frequencies will cancellation occur?

Solution
For the two signals to cancel $2\pi d/\lambda = n\pi$, where d is the path-length difference and n is an odd integer. Therefore

$$\frac{2d}{\lambda} = n$$

or $\quad 2d = n\lambda = \dfrac{nc}{f}$

and $\quad f = \dfrac{n \times 3 \times 10^8}{2 \times 75 \times 10^3} = 2n$ kHz.

Hence, the side-frequencies that cancel out are $f_C \pm 2000$ Hz, $f_C \pm 6000$ Hz, etc. (*Ans.*).

Frequency Diversity

Signals at different frequencies received by the same aerial very rarely fade simultaneously. This fact is used as the basis of a frequency-diversity system. A single aerial is connected to a number of radio receivers, each of which is tuned to a different frequency, whose outputs are commoned. The receiver circuitry is so arranged that the receiver that is instantaneously receiving the strongest input signal will provide the output signal. The obvious disadvantage of frequency diversity is its use of two, or more, frequencies at the same time.

Space Diversity

Signals at the same frequency that are received by two aerials sited several wavelengths apart rarely fade simultaneously. In a space-diversity system two, or three (but rarely more), aerials are sited some distance apart and are connected to two, or three, radio receivers. Each receiver is tuned to the same frequency and has a commoned output. As with frequency diversity the circuitry is arranged so that the receiver that is receiving the strongest signal supplies the output. The disadvantage of space diversity is the need for more than one aerial and the large site area required.

Example 8.5

At the distant end of an 8 MHz sky-wave radio link signals are received via two paths that are at angles of 12° and 24° to the ground. Calculate the optimum distance between the aerials in a two-aerial space-diversity system.

Solution

Consider Fig. 8.13. The wavefront of signal 1 arrives first at aerial A and then at aerial B, so that the signal at B lags the signal at A by angle

$$\phi_1 = \frac{2\pi d}{\lambda} \cos 12° \text{ radians.}$$

Similarly, for signal 2 the phase difference between the signals at aerials A and B is

$$\phi_2 = \frac{2\pi d}{\lambda} \cos 24° \text{ radians.}$$

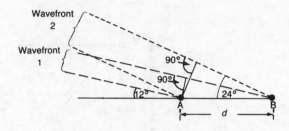

Fig. 8.13

For the optimum space diversity the difference between the two phase angles should be equal to π radians. Therefore

$$\frac{2\pi d}{\lambda}(\cos 12° - \cos 24°) = \pi$$

and $d = \dfrac{3 \times 10^8}{8 \times 10^6 \times 2\,(\cos 12° - \cos 24°)} = 290$ m.   (*Ans.*)

Space-wave Propagation

At frequencies in the v.h.f., u.h.f. and s.h.f. bands the main mode of propagation between two points on the surface of the earth is the *space wave*. Since the wavelength of the signal is small, both the transmitting and the receiving aerials can be mounted at a height of several wavelengths above ground. Figure 8.14 illustrates the principle of space-wave propagation. A radio wave travelling in the troposphere follows a slightly curved path because of tropospheric refraction, and this results in the radio horizon being more distant than the optical horizon. At distances less than the optical horizon reception is by means both of a direct wave and of a ground-reflected wave, but at greater distances, up to the radio horizon, only the direct wave is received. Some signals are also received at distances greater than the radio horizon because some diffraction takes place.

Figure 8.15 shows how, typically, the magnitude of the reflection coefficient $|\rho|$ of the earth may vary with the angle of incidence for both horizontally polarized and vertically polarized waves. For a horizontally polarized wave $|\rho|$ is always equal to unity, but for a vertically polarized wave $|\rho|$ varies considerably with the angle of incidence. The angle of the reflection coefficient $\angle\rho$ is always 180° for a horizontally polarized wave, but for a vertically polarized wave $\angle\rho$ varies from about 180° to about 10° as the angle of incidence is increased from zero. In practice, the angle of incidence is always

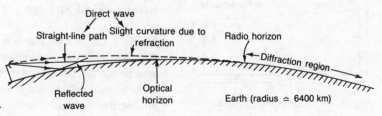

Fig. 8.14 Space-wave propagation.

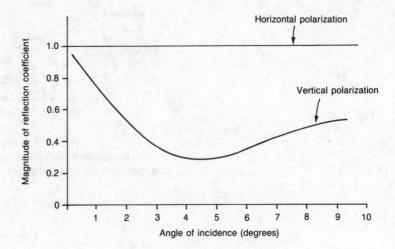

Fig. 8.15 Reflection coefficient of the earth.

small (and it is often called the *grazing angle*) and little error results if $\angle\rho$ is assumed to be 180°.

The relative merits of using horizontal, or vertical, polarization are as follows. The total field strength at the receive aerial is the resultant of the field strengths due to several components, including some diffracted energy. A wave that has been diffracted over a treeless hill will suffer less attenuation if it is vertically polarized. Conversely, if the hill is tree-covered a vertically polarized wave will be scattered to a greater extent and will thus suffer the greater attenuation. Reflected signals arrive at the receive aerial after reflection from the earth in front of the aerial, and from objects either side of the radio path. Reflecting objects in the vertical plane, such as hills, will produce a stronger reflected signal if the wave is vertically polarized.

In general, it is found that vertical polarization gives a larger received field strength at low heights above the ground but the probability of fading is greater. In hilly and/or wooded areas horizontal polarization is probably the better but vertical polarization is preferred for links that pass over flat countryside. For horizontal polarization the received field strength falls to zero at heights below about 4 metres. This does not matter for point-to-point links since the receive aerial is always mounted at greater heights than that, but it does mean that mobile land systems must employ vertical polarization.

Received Field Strength

Figure 8.16 shows a line-of-sight radio link; the *k*-factor is 4/3 so that the radio wave travels in a straight-line path and the earth is assumed to be flat. Two aerials, one at height h_t above ground, and the other at height h_r, are *D* kilometres apart. The total field strength produced at the receive aerial is the phasor sum of the individual field

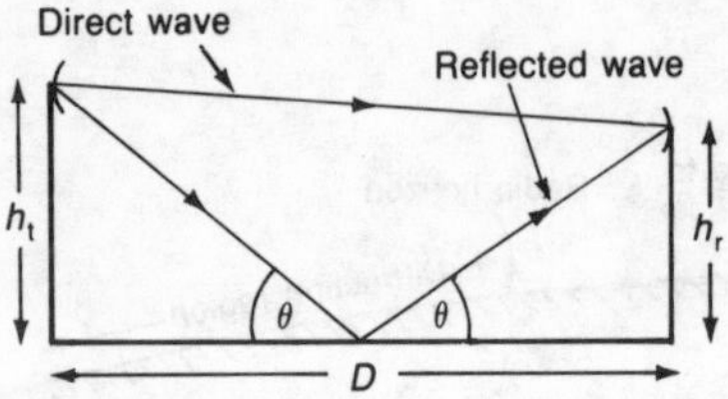

Fig. 8.16 A line-of-sight radio link.

strengths produced by the direct wave and by the reflected wave. The magnitudes of these waves are inversely proportional to the distance they have travelled. Since the extra length of the reflected path is negligible compared to the distance D between the two aerials any difference in amplitude due to this factor is negligibly small. Hence if $|\rho| = 1$, $|E_D| = |E_R| = E_1/D$; if $|\rho| \neq 1$, then $|E_R| = |\rho|E_1/D$. The amplitude of the resultant field strength will be a function of the phase difference between the direct and the reflected waves. This phase difference exists because of the angle of the ground-reflection coefficient and because of the difference in the direct and reflected path lengths. For small grazing angles the phase change upon reflection is approximately constant at $180°$. The phase difference ϕ due to the path length difference is $\phi = 2\pi/\lambda$ times that difference. The phasor diagram of the field strengths at the receive aerial is shown by Figs 8.17(a) and (b). Consider Fig. 8.17(a) in which $|\rho| = 1$ so that $|E_D| = |E_R|$; this figure has been re-drawn in Fig. 8.17(c) from which

$$AB = E_D \cos\left[\frac{\pi - \phi}{2}\right] = E_D \sin\left[\frac{\phi}{2}\right]$$

and $E_T = 2AB = 2E_D \sin\left[\frac{\phi}{2}\right] = \frac{2E_1}{D} \sin\left[\frac{\phi}{2}\right]$ (8.10)

Figure 8.17(b) is the phasor diagram when $|E_D| \neq |E_R|$; resolving E_R into its horizontal and vertical components gives the diagram shown in Fig. 8.17(d). From this

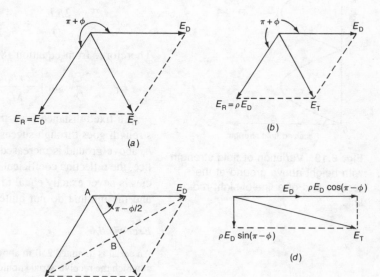

Fig. 8.17 Phasor diagrams of the received field strength in a line-of-sight link.

$$E_T = \sqrt{\{[(E_D + |\rho|E_D \cos(\pi - \phi)]^2 + |\rho|^2 E_D^2 \sin^2(\pi - \phi)\}}$$

$$= E_D\sqrt{[1 + |\rho|^2 + 2|\rho| \cos(\pi - \phi)]}$$

$$\text{or} \quad E_T = \frac{E_1}{D}\sqrt{[1 + |\rho|^2 - 2|\rho| \cos\phi]}. \tag{8.11}$$

If $|\rho| = 1$,

$$E_T = \frac{E_1}{D}\sqrt{[2(1 - \cos\phi)]} = \frac{E_1}{D}\sqrt{\left[4\sin^2\left(\frac{\phi}{2}\right)\right]}$$

$$= 2\frac{E_1}{D}\sin\left(\frac{\phi}{2}\right)$$

as before.

It is now necessary to determine the angle ϕ. Figure 8.18 is an extension of Fig. 8.16. From this figure

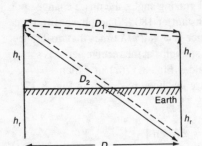

Fig. 8.18 Calculation of the angle ϕ.

$$D_1^2 = D^2 + (h_t - h_r)^2 \simeq D + \frac{(h_t - h_r)^2}{2D}$$

$$D_2^2 = D^2 + (h_t + h_r)^2 \simeq D + \frac{(h_t + h_r)^2}{2D}.$$

The difference between the lengths of the direct and the reflected paths is

$$D_2 - D_1 = \frac{(h_t + h_r)^2 - (h_t - h_r)^2}{2D} = \frac{2h_t h_r}{D}$$

and hence the phase angle

$$\phi = \frac{2\pi}{\lambda}\frac{2h_t h_r}{D} = \frac{4\pi h_t h_r}{\lambda D}.$$

Therefore, from equation (8.10),

$$E_T = \frac{2E_1}{D}\sin\left[\frac{2\pi h_t h_r}{\lambda D}\right]. \tag{8.12}$$

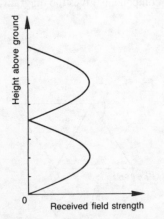

Fig. 8.19 Variation of field strength with height above ground at the receive end of a line-of-sight radio link.

At a fixed distance from the transmitting aerial the received field strength goes through successive maxima and minima as the height h_r above ground is increased. This is shown by Fig. 8.19. In practice, the reflection coefficient of the earth at u.h.f. and higher frequencies is never exactly equal to $1 \angle 180°$ because of surface roughness and the minima do not quite reach zero.

Example 8.6

An aerial is mounted 250 m above flat earth. Determine the minimum height at which the receive aerial should be mounted if it is to receive the maximum field strength. The distance between the aerials is 22 km and the frequency is 600 MHz.

Solution

From equation (8.12) the maximum field strength is obtained when

$$\frac{2\pi h_t h_r}{\lambda D} = \frac{n\pi}{2},$$

where n is an odd integer, or

$$h_r = \frac{n\lambda D}{4h_t}.$$

The minimum aerial height occurs when $n = 1$ and therefore

$$h_r = \frac{3 \times 10^8 \times 22 \times 10^3}{600 \times 10^6 \times 250 \times 4} = 11 \text{ m.} \quad (Ans.)$$

The distance D between the two aerials is always much larger than the heights h_t and h_r of the aerials. This means that $2\pi h_t h_r/\lambda D$ is a small angle and so equation (8.12) can be written as

$$E_T \simeq \frac{2E_1}{D}\frac{2\pi h_t h_r}{\lambda D}$$

or $\quad E_T = \frac{4\pi E_1 h_t h_r}{\lambda D^2}.$ \hfill (8.13)

This means that the field strength at a fixed height h_r above the ground is inversely proportional to the square of the distance from the transmitter, and directly proportional to frequency. As the distance is increased and nears the optical horizon the field strength tends to be equal to that produced by the direct wave alone and to be independent of frequency.

Curvature of the Earth

The assumption made, in deriving expressions for the received field strength, that the earth between the aerials is flat is, of course, not true. The error implicit in the assumption, however, is small unless the length of the link is approaching the optical horizon. If the curvature of the earth is to be taken into account a *divergence factor F* must be introduced that allows for the wave to be reflected from a curved surface. The divergence factor is employed by multiplying it and the reflection coefficient together, i.e. $E_R = \rho F E_D$.

Maximum Distance between Aerials

Because of the curvature of the earth there is a maximum distance from the transmitting aerial at which the receive aerial can be sited and still be able to receive the direct wave. This distance is known as the radio horizon. It is shown in Fig. 8.20 in which the effective radius of the earth is $kR = 8500$ km, h_t and h_r are heights of the transmit and receive aerials, and D_1 and D_2 are the distances from each aerial to the point of grazing incidence. At this point the direct

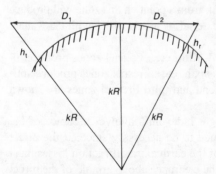

Fig. 8.20 Maximum distance between two aerials.

wave is tangential to the surface of the earth (when there will not be a reflected wave). From the figure

$$(h_t + kR)^2 = D_1^2 + (kR)^2$$

or $D_1 \simeq \sqrt{(2h_t kR)}$.

Similarly, $D_2 \simeq \sqrt{(2h_r kR)}$. The maximum distance between the aerials is

$$D_{max} = D_1 + D_2 = \sqrt{(2h_t kR)} + \sqrt{(2h_r kR)}. \tag{8.14}$$

If the heights of the two aerials are measured in metres and the distance D is in kilometres,

$$D_{max} = \sqrt{\left(\frac{2h_t}{1000} \times 8500\right)} + \sqrt{\left(\frac{2h_r}{1000} \times 8500\right)},$$

or $D_{max} = 4.13(\sqrt{h_t} + \sqrt{h_r})$. \tag{8.15}

Fresnel Zones

For a good, reliable signal to be received by an aerial there must be adequate clearance between the direct path and any obstacles. The necessary clearance is usually expressed in terms of *Fresnel zones*. A Fresnel zone is the locus of the points from which the sum of the distances to the transmitting and receiving aerials is equal to the direct distance between the aerials plus an integral number of half wavelengths.

The first Fresnel zone is the locus of the points from which the sum of the distances to each aerial is $\lambda/2$ longer than the direct path between the aerials. Since the ground-reflected wave experiences approximately 180° phase change upon any reflection that takes place at a point on the first Fresnel zone, it results in a signal which, at the receive aerial, is in phase with the direct wave.

The second Fresnel zone is the locus of the points from which the sum of the distances to each aerial is λ longer than the direct path between the aerials. Reflection from a point on this zone will produce a signal which, at the receive aerial, is in anti-phase with the wanted signal.

In similar fashion, the third, fifth, etc., Fresnel zones produce in-phase signals, and the fourth, sixth, etc., Fresnel zones produce anti-phase signals. The first, second and third Fresnel zones are shown by Fig. 8.21.

The design of a line-of-sight radio link involves a choice of the Fresnel zone clearance deemed to be necessary since, if the direct path is too near the surface of the earth extra diffraction losses must be expected. To avoid this, the clearance above ground of the direct ray should be equal to about 0.6 times the radius of the first Fresnel zone. Account must also be taken of possible sub-refraction effects

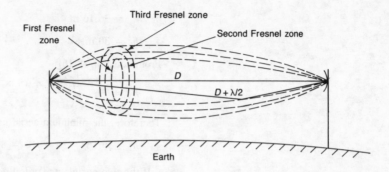

Fig. 8.21 The first, second and third Fresnel zones.

when the atmospheric conditions are such that the k-factor becomes smaller than unity. In the British Isles the k-factor rarely falls below 0.7 and so this is the value that is generally used in link design. Reflections from more distant objects tend to cancel out.

Radius of a Fresnel Zone

In Fig. 8.22 the distance $T-A-R$ is $\lambda/2$ longer than the distance $T-R$ so that the point A is on the first Fresnel zone. The radius of this zone is r and hence

$$(D_1^2 + r^2) = (D_2^2 + r^2) = D_1 + D_2 + \lambda/2$$

$$D_1\left(1 + \frac{r^2}{2D_1^2}\right) + D_2\left(1 + \frac{r^2}{2D_2^2}\right) = D_1 + D_2 + \lambda/2$$

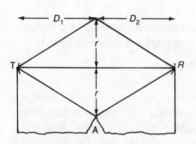

Fig. 8.22 Calculation of the radius of the first Fresnel zone.

$$\text{or} \quad r = \sqrt{\left(\frac{D_1 D_2 \lambda}{D_1 + D_2}\right)}. \tag{8.16}$$

The radius of any of the higher-order Fresnel zones is simply found by multiplying equation (8.16) by $\sqrt{n}$.

Example 8.7

A 3 GHz radio link has transmitting and receiving aerials at equal heights above the ground, and it is 36 km long. There is a 30 metre high obstacle midway between the two aerials. Determine the minimum height at which the aerials ought to be mounted for the radio path to be unobstructed. Assume the k-factor to be 0.7.

Solution

The minimum height of the two aerials must be equal to $h_1 + h_2 + h_3$, where h_1 is the height for grazing incidence, h_2 is the height of the obstacle, and h_3 is 0.6 times the radius of the first Fresnel zone.

$$\lambda = \frac{(3 \times 10^8)}{(3 \times 10^9)} = 0.1 \text{ m.}$$

From equation (8.14),

$$18 \times 10^3 = \sqrt{(2h_1 \times 0.7 \times 6400 \times 10^3)}$$

or $h_1 = 36.16$ m

$h_2 = 30$ m.

From equation (8.16),

$$h_3 = 0.6 \sqrt{\left(\frac{0.1 \times 18 \times 10^3}{2}\right)} = 18 \text{ m.}$$

Therefore, the minimum aerial height is

$$h = 36.16 + 30 + 18 = 84.16 \text{ m.} \quad (Ans.)$$

If the two aerials are not at equal heights and/or the obstacle is not at mid-path the problem is more complex. It is best approached by assuming the earth between the aerials to be flat and then increasing the effective height of the obstacle(s) by the amount necessary to account for the curvature of the earth. The effective increase in height h_{inc} of the obstacle is given by equation (8.17), i.e.

$$h_{\text{inc}} = \frac{D_1 D_2}{2kR}, \tag{8.17}$$

where D_1 is the distance from the transmitting aerial to the obstacle and D_2 is the distance from the obstacle to the receive aerial.

Example 8.8

A 3 GHz signal is transmitted from an 80 m high aerial towards a receiving aerial that is 39 km away. A 50 m high obstacle is 25 m from the transmitter. Calculate the necessary minimum height of the receive aerial. Assume the k-factor to be 0.7 and allow a clearance equal to 0.6 times the radius of the first Fresnel zone.

Solution

When flat earth is assumed the obstacle must be given an effective height of

$$\frac{25 \times 14 \times 10^6}{2 \times 0.7 \times 6400 \times 10^3} = 39.1 + 50 \simeq 89 \text{ m.}$$

The first Fresnel zone clearance is

$$0.6 \sqrt{\left(\frac{0.1 \times 25 \times 14 \times 10^6}{39 \times 10^3}\right)} = 18 \text{ m.}$$

From Fig. 8.23

$$\tan \theta = \frac{(89 + 18) - 80}{25 \times 10^3} = \frac{h_r - 80}{39 \times 10^3}$$

or $h_r = 122$ m.   (Ans.)

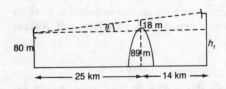

Fig. 8.23

Height Diversity

If the radio path crosses over a wide stretch of water, such as a tidal estuary, or the sea, the geometry of the link may vary with the tide.

It is often necessary to employ some form of diversity reception in order to obtain a reliable system. To economize with the usage of frequency it is usual to employ *height diversity* in which two receiving aerials are used, one mounted above the other. For the optimum results the lower aerial should be mounted at the height corresponding to the first maximum in the field strength/height characteristic (Fig. 8.19), and the other aerial at the height of the null immediately above.

Scatter Propagation

A tropospheric-scatter radio link operates with its distant terminal well beyond the radio horizon. A large amount of radio-frequency energy is radiated, by a highly directive aerial, towards the horizon. A very small proportion of the radiated energy is *forward scattered* by the troposphere and is directed downwards towards the receive aerial. Most of the transmitted energy continues upwards, passes through the ionosphere and is radiated into space. Figure 8.24 shows the path geometry of a tropospheric-scatter radio link. The solid angles formed by the narrow radiation patterns of the two aerials intercept one another to form a common volume that is known as the *scatter volume*. The scatter volume is typically only one, or two, kilometres above earth and it is here that the useful energy is returned to earth. Both the transmitter *launch angle* and the *scatter angle* should be small, usually less than about 4°.

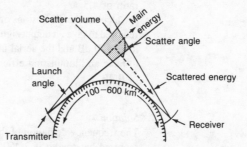

Fig. 8.24 Tropospheric scatter link.

Tropospheric-scatter propagation is possible at most frequencies higher than 500 MHz, but it is employed mostly at frequencies in the region of 900 MHz, 2 GHz and 5 GHz. It provides a bandwidth of several megahertz and is employed to carry wideband telephony systems, with between 38 and 132 channels. The main application for a tropospheric-scatter system is for communication over hostile terrain, such as sea or mountains, where a communications satellite system would not be economically viable. It is the principal broadband communication technology employed for communication between the mainland and the North Sea oil rigs; these operate at 2.25 GHz with a transmitted power of 1 kW.

The path attenuation between transmitter and receiver is much greater than the free-space attenuation (given by equation (6.28)), by an *over-the-horizon loss*, or *scatter loss*. The scatter loss accounts for the fact that only a small fraction of the transmitted power arrives at the receive aerial. Since this loss is high, typically 70 dB or more, a tropospheric-scatter system must be provided with high-power transmitters, high-gain aerials, and low-noise factor, highly sensitive radio receivers. The received signal is further reduced by the *aperture-to-coupling loss*. This expresses the loss of gain of a parabolic dish aerial when it receives a scattered signal and it is typically about 10 dB; it occurs because the received signal is made up of a large number of components bearing a random phase and amplitude relationship with one another. The scatter loss is continually varying and this leads to substantial fading of the received signal. To counter this some kind of diversity reception is usual and perhaps the most commonly employed is *quadruple space diversity*. This involves the use of four aerials: two at each end of a route. The two transmitting aerials radiate energy simultaneously, with their signals polarized at right angles to one another. The two receive aerials pick up signals in both the horizontal and the vertical planes and feed receivers whose outputs are commoned.

Example 8.9

A 200 km tropospheric-scatter system operates at 2 GHz with a transmitted power of 1 kW and a receiver of noise factor 5 dB. The system uses frequency modulation with a maximum modulating frequency of 100 kHz and a deviation ratio of 4.5.

Calculate the dimensions of the identical transmitting and receiving aerials required to give an output signal-to-noise ratio of 60 dB. The over-the-horizon loss is 70 dB and the aerial noise temperature is 290 K. Assume the aerials to have an illumination efficiency of 60% and an aperture-to-medium loss of 9 dB.

Solution

The frequency deviation of the signal is $4.5 \times 100 = 450$ kHz. Hence the required bandwidth is $2(450 + 100) \times 10^3 = 1.1$ MHz.

$$60 \text{ dB} = P_r \text{ dBW} + 20 \log_{10} (\sqrt{3}D) \text{ dB} - kTB \text{ dBW} - F \text{ dB}$$

$$= P_r \text{ dBW} + 17.84 \text{ dB} - 143.6 \text{ dBW} + 5 \text{ dB}$$

$$= -96.4 \text{ dBW} = 2.29 \times 10^{-10} \text{ W}.$$

From equation (6.28)

$$2.29 \times 10^{-10} = G_t G_r \times 1000 \left(\frac{3 \times 10^8}{2 \times 10^9 \times 4\pi \times 200 \times 10^3} \right)^2$$

$$\times 10^{-7} \times 0.125$$

$$G_t G_r = G^2 = \frac{2.29 \times 10^{-10}}{4.45 \times 10^{-20}} = 5.143 \times 10^9$$

$$G_t = G_r = 71\,715 \quad \text{or} \quad 48.6 \text{ dB}.$$

Therefore,

$$71\ 715 = 0.6\left(\frac{\pi D}{\lambda}\right)^2 = 0.6\left(\frac{\pi D}{0.15}\right)^2$$

or $D \simeq 16.5$ m. (Ans.)

Propagation via a Communications Satellite

The power received by the ground station of a communications satellite link can be determined using equation (6.28). The term $(4\pi D/\lambda)^2$ is known as the *transmission loss* and it accounts for the way in which the radio wave diverges as it travels. This relationship can be alternatively expressed as

$$P_r = \frac{\text{effective radiated power} \times \text{gain of receive aerial}}{\text{transmission loss}}.$$

(8.18)

The received power is often referred to as the carrier power.

In practice, further losses are experienced because of climatic conditions such as rain, snow, etc.

Example 8.10

A communications satellite is 40 000 km from a point on the surface of the earth and it transmits a power of 2 W from an aerial of 20 dB gain. Calculate the power received at the earth station by an aerial of effective aperture 10 m if the frequency is 11 GHz.

Solution
Method (*a*)

$$\lambda = \frac{3 \times 10^8}{11 \times 10^9} = 0.027 \text{ m}.$$

The flux density at the receive aerial is

$$P_a = \frac{P_t G_t}{4\pi D^2} = \frac{2 \times 100}{4\pi(40 \times 10^6)^2} = 9.95 \times 10^{-15} \text{ W/m}.$$

The received power

$$P_r = P_a A_e = 9.95 \times 10^{-15} \times 10 = 9.95 \times 10^{-14} \text{ W}. \quad (Ans.)$$

Method (*b*)

$$\text{Transmission loss} = \left(\frac{4\pi \times 40 \times 10^6}{0.027}\right)^2 = 3.466 \times 10^{20}.$$

$$\text{Gain of the receive aerial} = \frac{4\pi A_e}{\lambda^2} = \frac{4\pi \times 10}{0.027^2} = 172\ 378.$$

Therefore

$$\text{the received power} = \frac{2 \times 100 \times 173\ 278}{3.466 \times 10^{20}}$$

$$= 9.95 \times 10^{-14} \text{ W}. \quad (Ans.)$$

Figure-of-Merit or G/T Ratio

In a communications satellite system the received power levels are always *very* small because of the extremely large distances involved. This means that the noise generated within the receiving system must be reduced to the minimum possible value in order to achieve a satisfactory carrier-to-noise ratio. There are two main requirements that must be satisfied to achieve this. First, the bandwidth B of the receiver must be as narrow as possible, and second, the system noise temperature must be as low as possible.

If the system noise temperature (p. 109) is T_S then the noise power P_{in} at the input to the demodulator is given by $P_{in} = GkT_S B$, where G is the gain of the receiver from the r.f. input to the demodulator input. Since the received carrier power is P_r watts the carrier-to-noise ratio at the demodulator input is

$$\frac{C}{N} = \frac{GP_r}{GkT_S B} = \frac{P_t G_t G_r}{kT_S B} \left(\frac{\lambda}{4\pi D} \right)^2$$

$$= \frac{P_t G_t}{kB} \left(\frac{\lambda}{4\pi D} \right)^2 \frac{G_r}{T_S}. \tag{8.19}$$

The carrier-to-noise ratio is proportional to the ratio (gain of receive aerial)/(receiver system noise temperature). Since, for a given system $\dfrac{P_t G_t}{kB} \left(\dfrac{\lambda}{4\pi D} \right)^2$ is a constant, the performance of a communications satellite system can be improved by increasing its G/T ratio. A typical figure for the G/T ratio, or *figure-of-merit*, is 40 dB^{-1}K.

Example 8.11

The parabolic dish aerial used by the ground station of a communications satellite link has a diameter of 28 m and an illumination efficiency of 70%. The link operates at a frequency of 4 GHz. Calculate the G/T ratio of the system if the system noise temperature is 80 K.

Solution

$$\lambda = \frac{3 \times 10^8}{4 \times 10^9} = 0.075 \text{ m}.$$

The gain of the receive aerial is

$$\frac{\eta 4\pi A_e}{\lambda^2} = \frac{0.7 \times 4\pi^2 \times 14^2}{0.075^2} = 962\,922.5 \quad \text{or} \quad 59.8 \text{ dB}.$$

Also, $T = 80 \text{ K} = 19 \text{ dB}^{-1}\text{K}$. Therefore,

$$G/T \text{ ratio} = 59.8 - 19 = 40.8 \text{ dB}^{-1}\text{K}. \quad (Ans.)$$

9 Communication Radio Receivers

The function of a communication radio receiver is to select the wanted signal present at the receive aerial, which may be of very small amplitude, from the background noise and to reject a large number of, possibly stronger, unwanted signals. The radio receiver must then amplify and demodulate the received signal to provide an output baseband signal with at least the minimum required signal-to-noise ratio. The main problem is not amplifying the received signal to the wanted level but overcoming the adverse effects of both the received noise and internally generated noise and interference.

Most m.f./h.f. communication radio receivers are able to receive different kinds of signal, such as d.s.b. amplitude modulation, s.s.b.s.c., c.w. and data, and in some cases frequency modulation also. Receivers designed for use in the v.h.f./u.h.f. bands usually receive either amplitude-modulation or frequency-modulation signals. Modern communication receivers are mostly of the double superheterodyne type; less often single-, or triple-superheterodyne receivers are employed. There are two main reasons for this trend: first, the first intermediate frequency (i.f.) can be high, giving a wide separation between the wanted signal frequency and the image-channel signal frequency; and, second, the second i.f. can be low, making good adjacent-channel selectivity easier to obtain. Since many applications for communication radio receivers require the receiver to be remotely controlled and/or able to change frequency rapidly and accurately, broadband front-ends and frequency synthesis are commonly employed.

Double-Superheterodyne Radio Receivers

The block diagram of a typical h.f. communication radio receiver is shown by Fig. 9.1. The first i.f. of 45 MHz is above the 2−30 MHz tuning range of the receiver. The second i.f. is at the (more or less) standard value of 1.4 MHz. The first, and the second, local oscillator frequencies which must be supplied to the two mixers are both derived from a frequency synthesizer. Signals picked up by the aerial are passed through the 30 MHz low-pass filter to remove all signals at frequencies above the tuning range, but especially any signals at, or near, the image-channel frequency, and the first intermediate frequency. The r.f. stage should transfer the maximum r.f. power from the aerial to the receiver and this means that it should be matched to the aerial. This, however, is not always possible since the receiver may be used in conjunction with more than one type of aerial. In some

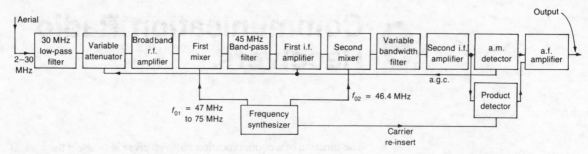

Fig. 9.1 An h.f. communication radio receiver.

older types of h.f. receiver the r.f. stage consists of one, or more, r.f. amplifier stages that are ganged with the first local oscillator. Some modern receivers employ a bank of sub-octave band-pass filters any one of which may be switched into circuit, quite likely by microprocessor-controlled digital circuitry.

The signals are then passed through an attenuator whose attenuation is varied by the automatic gain-control (a.g.c.) system. The r.f. attenuator is followed by a broadband amplifier and the first mixer (an r.f. amplifier is not provided in some receivers). Besides amplifying the received signal and improving the noise factor of the receiver, the amplifier also isolates the first mixer from the aerial and helps to prevent local oscillator radiation from the aerial. The first mixer mixes the amplified signal with the first oscillator frequency and generates a component at their difference frequency $f_{01} - f_S =$ 45 MHz. The frequency supplied by the frequency synthesizer must therefore be variable from 47 MHz to 75 MHz. The difference-frequency component is then selected by the first i.f. band-pass filter and it is then amplified. The selected signal is then mixed with the fixed second local oscillator frequency of 46.4 MHz to shift it to the second intermediate frequency of 1.4 MHz. The wanted 1.4 MHz component of the output of the second mixer is selected by one of a number of band-pass filters. Each filter has a centre frequency of 1.4 MHz but a bandwidth appropriate for the different kinds of signals handled by the receiver, e.g. 8 kHz for a.m., 2.4 kHz for s.s.b., 1 kHz or 400 Hz for c.w. After filtering, the signal is passed to either one of the two detectors. The a.m. detector demodulates d.s.b.a.m. signals and also generates the a.g.c. voltage. The product detector is used to demodulate all other kinds of signal. Finally, the demodulated signal is amplified by the audio amplifier.

The quality of the output signal is affected by a number of factors, amongst which are: (a) adjacent-channel, co-channel and image-channel signals, (b) reciprocal mixing, and (c) cross-modulation and intermodulation. In the design of a modern communication receiver every effort is made to minimize the adverse effects of these factors; this means that particular attention is paid to r.f. linearity and filtering and to i.f. selectivity. Increasingly, modern designs make full use of readily available ICs to provide such circuit features as amplification, detection, frequency synthesis and digital tuning and displays.

The block diagram of a v.h.f./u.h.f. communication radio receiver follows a broadly similar scheme to Fig. 9.1. Typically, the receiver might cover the frequency band 20 MHz to 470 MHz with a first intermediate frequency of 515 MHz. The second intermediate frequency is nearly always 10.7 MHz. The received signals may be either amplitude, or frequency modulated.

Choice of the Local Oscillator Frequency

The first intermediate frequency of a communication radio receiver is equal to the difference between the first local oscillator frequency f_{01} and the wanted signal frequency f_S. It is usual to make the oscillator frequency higher than the signal frequency since the ratio $f_{01(max)}/f_{01(min)}$ is then smaller. Assuming this, there are two possibilities: *either* the wanted signal can be shifted to a lower intermediate frequency, known as *down-conversion*, *or* it can be shifted to a higher intermediate frequency, known as *up-conversion*. For modern h.f. communication radio receivers up-conversion is the more common technique since it ensures that all possible image-channel signals lie above the tuning range of the receiver. The image-channel signal can then be suppressed by a low-pass filter whose cut-off frequency is equal to the highest frequency to which the receiver can tune. A further advantage of up-conversion is that the ratio $f_{01(max)}/f_{01(min)}$ is smaller than for down-conversion.

Image-channel Interference

When a superheterodyne radio receiver has been tuned to a frequency f_S there will always be another frequency, known as the *image-channel frequency* f_{im}, that will also produce the first intermediate frequency f_{if1} if it is allowed to reach the first mixer. Now,

$$f_{if1} = f_{01} - f_S = f_{im} - f_{01},$$

$$f_{if1} = f_{im} - (f_{if1} + f_S)$$

or $f_{im} = f_S + 2f_{if1}.$

Thus, the image-channel signal is separated from the wanted signal by a frequency gap equal to twice the first intermediate frequency. The r.f. stage must include sufficient selectivity to stop the image-channel signal reaching the first mixer.

The frequency of the image-channel signal will vary as the receiver is tuned to receive different signals. The r.f. stage must therefore be able to reject image-channel signals at different frequencies. Traditionally, the necessary front-end selectivity has been provided by variable-tuned resonant circuits ganged with the first local oscillator. Increasingly, modern communication receivers either employ a single low-pass filter (as in Fig. 9.1), or a bank of sub-octave band-pass filters.

Any vestige of the image-channel signal that reaches the first mixer will cause an unwanted signal to appear at the first intermediate frequency. This signal cannot be rejected by the first i.f. filter and so it will produce crosstalk at the output of the receiver. The *image-response ratio* is the ratio, in decibels, of the r.f. input voltages at the wanted signal frequency, and at the image-channel frequency, that are necessary to produce the same audio output power. Typically, an h.f. communication receiver might have an image-response ratio of 100 dB and a v.h.f./u.h.f. receiver of some 70 to 90 dB.

Choice of the Intermediate Frequencies

The main factors to be considered when choosing the intermediate frequencies for a communication receiver are: (*a*) interference signals, (*b*) adjacent-channel selectivity, (*c*) i.f. breakthrough, and (*d*) the availability of crystal and SAW filters.

The first intermediate frequency should not be in, or near, the tuning range of the receiver. Then the front end of the receiver can include filtering to prevent any signals at, or near, the first intermediate frequency reaching the first mixer. Usually, the first i.f. stage is isolated from the r.f. input by at least 80 dB, but preferably 120 dB.

The smaller the frequency separation between the wanted signal and the image-channel signal the harder it is to achieve adequate suppression of the image-channel signal. This factor requires the first intermediate frequency to be as high as possible. Conversely, good adjacent-channel selectivity is easier to obtain if the second intermediate frequency is low. The actual frequencies chosen are decided by the frequencies at which crystal and SAW filters are readily available.

Co-channel Interference

Co-channel interference is caused by an unwanted signal at the same frequency as the wanted signal. Clearly, it cannot be eliminated either by filtering or by the selectivity of the receiver. In a v.h.f./u.h.f. frequency-modulation receiver co-channel interference is not important since it is eliminated by the capture effect.

Intermodulation

When two, or more, signals at frequencies f_1 and f_2 are applied to a non-linear characteristic they will generate *intermodulation products*. These may be either *second-order* $f_1 \pm f_2$ products or *third-order* $2f_1 \pm f_2$, $2f_2 \pm f_1$, products. The more important of these are the third-order products since they tend to have frequencies that are within the passband of the first i.f. stage. For example, suppose that a wanted signal at 3.01 MHz is accompanied by two unwanted signals at 3.02 MHz and 3.03 MHz. The third-order intermodulation product $(2 \times 3.02) - 3.03$ is at the same frequency as the wanted signal.

A non-linear device has a transfer characteristic of the form

$$v = a + bv + cv^2 + dv^3 + \ldots, \tag{9.1}$$

where a, b, c and d are constants.

The intermodulation products are generated by both the square and the cube terms. Consider the square term, with inputs $V_1 \cos \omega_1 t$ and $V_2 \cos \omega_2 t$,

$$
\begin{aligned}
v_{\text{out}} &= c(V_1 \cos \omega_1 t + V_2 \cos \omega_2 t)^2 \\
&= cV_1^2 \cos^2 \omega_1 t + cV_2^2 \cos^2 \omega_2 t \\
&\quad + 2cV_1 V_2 \cos \omega_1 t \cos \omega_2 t. \tag{9.2}
\end{aligned}
$$

The first two terms give the second harmonics of the two input signals but these will not pass through the i.f. filter. The third term can be expanded to give

$$cV_1 V_2[\cos (\omega_1 - \omega_2)t + \cos (\omega_1 + \omega_2)t]$$

and this shows that the output voltage contains components at frequencies $f_1 \pm f_2$. These components are known as the *second-order intermodulation products*.

In general, second-order intermodulation products fall outside of the passband of the first i.f. filter and are suppressed. If, however,

$$f_2 = f_1 + \frac{f_{\text{if1}}}{2}$$

then

$$f_2 - f_1 = \frac{f_{\text{if1}}}{2}$$

and the second harmonic of this is equal to the first intermediate frequency. This is known as *half-i.f.*, or *repeat-spot*, interference. Alternatively, if the receiver is tuned to frequency f_S and there is an unwanted signal at frequency $f_S - f_{\text{if1}}/2$, the second harmonic of the unwanted signal will mix with the second harmonic of the local oscillator frequency to give the intermediate frequency.

Third-order intermodulation products are generated by the cubic term in equation (9.1). Thus

$$
\begin{aligned}
v_{\text{out}} &= d(V_1 \cos \omega_1 t + V_2 \cos \omega_2 t)^3 \\
&= d(V_1^3 \cos^3 \omega_1 t + V_2^3 \cos^3 \omega_2 t \\
&\quad + 3V_1^2 V_2 \cos^2 \omega_1 t \cos \omega_2 t \\
&\quad + 3V_1 V_2^2 \cos \omega_1 t \cos^2 \omega_2 t)
\end{aligned}
$$

$$= \frac{dV_1^3}{4} (3 \cos \omega_1 t + \cos 3\omega_1 t)$$

$$+ \frac{dV_2^3}{4} (3 \cos \omega_2 t + \cos 3\omega_2 t)$$

$$+ dV_1^2 V_2 [\tfrac{3}{2} \cos \omega_2 t + \tfrac{3}{4} \cos (2\omega_1 + \omega_2)t$$

$$+ \tfrac{3}{4} \cos (2\omega_1 - \omega_2)t]$$

$$+ dV_1 V_2^2 [\tfrac{3}{2} \cos \omega_1 t + \tfrac{3}{4} \cos (2\omega_2 + \omega_1)t$$

$$+ \tfrac{3}{4} \cos (2\omega_2 - \omega_1)t]. \tag{9.3}$$

Either of the $2\omega_1 - \omega_2$ or $2\omega_2 - \omega_1$ components may be at such a frequency that it falls within the passband of the first i.f. filter. Higher, even-order products are generally out-of-band while higher order, odd-order products are usually of negligible amplitude.

The *third-order intermodulation level* is the level, in decibels, relative to 1 μV (dBμV) of two unwanted signals, respectively 10 kHz and 20 kHz off-tune that generate an unwanted third-order output equivalent to that produced by a wanted signal at 0 dBμV e.m.f. A good communication receiver would have a level of about 85 dBμV.

To reduce intermodulation, a pre-selector stage, such as a bank of sub-octave filters, is often employed before the first non-linear stage. Alternatively (or in addition), an r.f. attenuator will reduce the third-order intermodulation level by three times the r.f. attenuation in dB. If, for example, a 6 dB attenuator is fitted it will reduce each interfering signal by 6 dB but, since the amplitude of the third-order products is proportional to $V_1^2 V_2$, or $V_1 V_2^2$, it will reduce the intermodulation level by 18 dB. Hence, the use of a 6 dB r.f. attenuator would increase the signal-to-intermodulation level by $18 - 6$ or 12 dB.

Intercept Point

Reference to equations (9.2) and (9.3) shows that if the amplitudes of the two unwanted signals are equal to one another, i.e. $V_1 = V_2 = V$, the amplitudes V_{IP} of the second-order and the third-order intermodulation products are given by

$$V_{IP} = k_n V^n, \tag{9.4}$$

where k_n is the nth-order constant, and n is the order of the intermodulation products.

This means that whereas the output voltage due to the wanted signal increases in direct proportion to its input voltage V the intermodulation output voltage increases in proportion to V^n. There must therefore be a level at which the two output voltages are equal to one another. This voltage, V_{PN}, is known as the *nth-order intercept point*. The intercept point is a purely theoretical level because the amplifier,

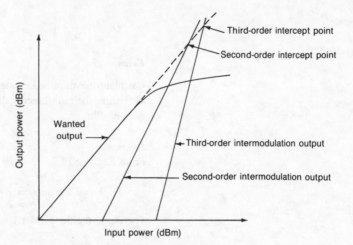

Fig. 9.2 Second- and third-order intercept points.

or mixer, would saturate before the level is reached. The intercept point is determined by extending, on a graph of output power in dBm plotted against input power in dBm, the wanted signal and the nth-order intermodulation product outputs to the point where they intercept each other. Figure 9.2 shows how output power varies with increase in the wanted signal, and for both the second-order and third-order intermodulation products. Since both axes are logarithmic the second-order plot has twice the slope, and the third-order plot has three times the slope, of the wanted signal output plot. The points at which the curves cross are known as the *second-order intercept point* and the *third-order intercept point*. The third-order intercept point is a measure of the ability of a receiver to reject large-amplitude unwanted signals whose frequency is near to that of the wanted signal. The third-order intercept point is primarily determined by the linearity of the first mixer and it should be as large as possible.

For the third-order intermodulation products $n = 3$ and equation (9.4) can be written in the form

$$P_{3IP} = \left[k_n \left(\frac{V}{2} \right)^3 \right]^2 = (k_n P)^3, \tag{9.5}$$

where $P = V^2/2$ and is the power due to one signal on its own. This means that the intermodulation power is proportional to the cube of the input power, i.e. a 1 dB increase in the input power gives a 3 dB increase in the third-order intermodulation output power.

The ratio D of the intermodulation power to the wanted signal power is $D = P_{3IP}/P_{out}$. P_{3IP} is proportional to P^3 and P_{out} is proportional to P. Hence

$$D = (KP)^2. \tag{9.6}$$

At the third-order intercept point $P_{3IP} = P_{out}$ and so D is equal to unity. Equation (9.6) becomes $1 = (KP)^2 = (KP_{3IP})^2$, or $K = P_{3IP}^{-1}$. Consequently, equation (9.6) can be written as

$$D = \left[\frac{P}{P_{3IP}} \right]^2. \tag{9.7}$$

Example 9.1

Calculate the third-order intercept of a receiver if, for an input power of 0 dBm, the ratio (intermodulation power)/(wanted power) at the output is −40 dBm.

Solution
From equation (9.7)

$$-40 = 2 \log_{10} \left[\frac{P}{P_{3IP}} \right] = 2 \times (0 - P_{3IP}) = -2P_{3IP}.$$

Therefore, the third-order intercept point is +20 dBm.   (*Ans.*)

Note from this example that for a 0 dBm two-tone signal the third-order intercept point is equal to $-\frac{1}{2}$ times the magnitude of the third-order intermodulation products.

When the two signals are not at 0 dBm they should both be normalized to 0 dBm. The third-order intermodulation level then increases by 3 dB for every 1 dB increase in the two-tone signal level. If the two signals are at different levels then subtract one third of their level difference in dBm from the larger level and take the result as being their common level.

Example 9.2

When two 0 dBm tones are applied to a mixer the level of the third-order intermodulation products is −60 dBm. The mixer has a conversion loss of 6 dB. Calculate, in dBm, the third-order intermodulation output power when the level of the two tones is (*a*) −10 dBm, (*b*) +10 dBm, (*c*) +30 dBm, and (*d*) +20 dBm and +11 dBm.

Solution
The level of the wanted signal at the mixer output is −6 dBm.

(*a*) The third-order output power is $(-60) + 3 \times (-10) = -90$ dBm.
(*Ans.*)

(*b*) The third-order output power is $(-60) + 3 \times 10 = -30$ dBm.
(*Ans.*)

(*c*) The third-order output power is $(-60) + 3 \times 30 = +30$ dBm.
(*Ans.*)

This answer means that the third-order output level is equal to the input level so this is the third-order intercept point.

(*d*) The equivalent input level is

$$20 - \frac{(20 - 11)}{3} = +17 \text{ dBm.}$$

The third-order output power is $(-60) + 3 \times 17 = -9$ dBm.
(*Ans.*)

Typical figures for an m.f./h.f. communication receiver are +15 dBm second-order and +20 dBm third-order intercept points. For a v.h.f./u.h.f. receiver typical figures are second-order intercept point +5 dBm up to 470 MHz, and 0 dBm from 470 MHz to 1.1 GHz.

Dynamic Range

The dynamic range of a radio receiver is the range of input levels that produce output powers lying in between the *noise floor* of the receiver and the input level that makes the *total* intermodulation product power equal to the noise floor. The dynamic range is given by

$$\text{dynamic range} = \tfrac{2}{3}[P_{3IP} - \text{noise floor}]$$

$$= \tfrac{2}{3}[P_{3IP} - FkT_0B \text{ (dB)}]. \qquad (9.8)$$

Example 9.3

Calculate the dynamic range of a receiver that has a third-order intercept point of +20 dBm, a noise factor of 6 dB and a bandwidth of 8 kHz. $kT = -174$ dBm.

Solution
From equation (9.8)

$$\text{dynamic range} = \tfrac{2}{3}[20 + 174 - 6 - 10_{10} \log 8000] \simeq 99 \text{ dB.}$$
$$(Ans.)$$

Reciprocal Mixing

When a large-amplitude off-tune signal appears at the input to the first mixer it will mix with the noise sidebands of the first local oscillator and may produce in-band noise. This process is known as *reciprocal mixing* and it is illustrated by Fig. 9.3 in which both the

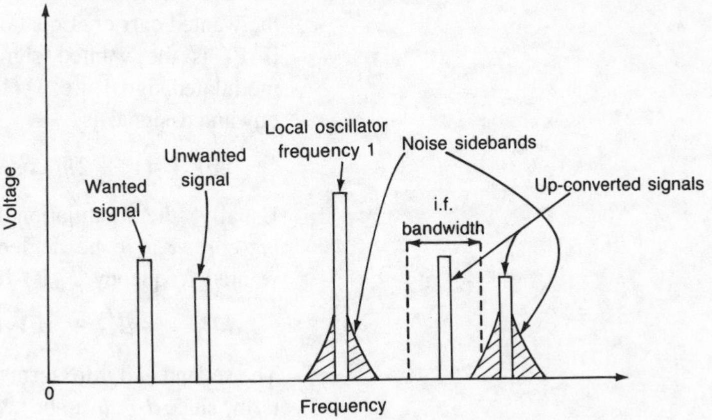

Fig. 9.3 Reciprocal mixing.

wanted signal and the unwanted signal are assumed to occupy narrow bandwidths that are much less than the first i.f. bandwidth. The first local oscillator has upper, and lower noise sidebands, sometimes known as *phase noise*, and these produce unwanted signals which lie within the passband of the first i.f. filter. Reciprocal mixing is defined as the amount of noise introduced by a 20 kHz off-tune signal that will produce an output equivalent to that produced by the wanted signal when its voltage is 1 μV e.m.f. Suppose, for example, that the frequency of the wanted signal is 12 MHz and that the first intermediate frequency is 45 MHz. The first local oscillator frequency is then 57 MHz. If there is a 45.01 MHz, 3 kHz slice of oscillator noise 93 dB down on the oscillator voltage then an unwanted signal at 12.01 MHz would be converted to a 3 kHz noise band at 45 MHz and spuriously received.

The reciprocal-mixing performance of a receiver affects its ability to reject off-tune signals and it means that the effective selectivity of the receiver is not as good as the selectivity defined by the i.f. filters. The effect of reciprocal mixing on selectivity is shown by Fig. 9.4; clearly, the selectivity characteristic is widened.

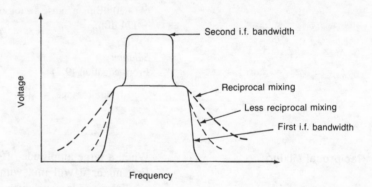

Fig. 9.4 Effect of reciprocal mixing upon selectivity.

Cross Modulation

Cross modulation is the transfer of the amplitude modulation of an unwanted carrier, that appears at the input to the first mixer, on to the wanted carrier. Equation (9.3) contains the term $\frac{3}{2}dV_1V_2^2 \cos \omega_1 t$; if V_1 is the wanted signal and V_2 is the unwanted amplitude-modulated signal, i.e. $V_2(1 + m \cos \omega_m t)$, then the amplitude of the unwanted signal is

$$\tfrac{3}{2}dV_1V_2^2(1 + 2m \cos \omega_m t + m^2 \cos^2 \omega_m t). \tag{9.9}$$

Usually, the modulation factor m is small enough for the term $m^2 \cos^2 \omega_m t$ to be neglected. Then the total output signal at the wanted frequency ω_1 is, from equations (9.1), (9.3) and (9.9)

$$bV_1 + \tfrac{3}{4}dV_1^3 + \tfrac{3}{2}dV_1V_2^2 + 3dV_1V_2^2m \cos \omega_m t.$$

The second and third terms are small enough compared with the first term, since $d \ll b$, to be neglected and hence the output at the wanted frequency ω_1 consists of the wanted component bV_1 *plus* an unwanted cross-modulation component $3dV_1V_2^2m \cos \omega_m t$. The signal-to-cross-modulation ratio SCMR is

$$\text{SCMR} = 20 \log_{10}\left(\frac{3dV_1V_2^2m}{bV_1}\right) \text{ dB}$$

$$= 20 \log_{10}\left(\frac{3dV_2^2m}{b}\right) \text{ dB}. \tag{9.10}$$

The cross-modulation factor is the ratio of the modulation factors of the superimposed modulation and of the unwanted signal, i.e.

$$\text{cross-modulation factor} = \frac{3dV_2^2m}{bm} = \frac{3dV_2^2}{b}. \tag{9.11}$$

Cross modulation is specified as the level of the 20 kHz off-tune signal, 30% modulated at 1 KHz, that produces an output signal 30 dB down on the level produced by the wanted signal, 30% modulated at 1 kHz, at 60 dBμV e.m.f.

Since both cross modulation and third-order intermodulation arise because of the cubic term in a non-linear characteristic it is to be expected that there is a relationship between them. Approximately, this relationship is

$$V_{\text{CM}}(\text{dB}\mu\text{V}) = \tfrac{3}{2}V_{3\text{IP}}(\text{dB}\mu\text{V}) - 6(\text{dB}) - \frac{\text{SCMR}}{2} \text{ dB}. \tag{9.12}$$

Blocking

Blocking is an effect in which the gain of a radio receiver is reduced when a large-amplitude unwanted signal either overloads a stage, or excessively operates the receiver's a.g.c. system. When blocking occurs the level of the wanted output signal falls each time the interfering signal is received. The *blocking ratio* is the ratio

(response to a signal at one frequency when there is a simultaneous excitation at another frequency)/(response to the one signal above). $\tag{9.13}$

The blocking ratio of a receiver depends both upon the magnitudes of the two signals and on their frequency difference. It is specified as the level of an interfering 20 kHz off-tune signal that gives a change in the wanted output signal of 3 dB, with the a.g.c. system of the receiver inoperative.

If the level of the two signals, in dBμV, to give 1 μV intermodulation product is $V_{3\text{IP}}$ then, approximately, the voltage V_B of the blocking signal is

$$V_B = \tfrac{3}{2}V_{3\text{IP}}(\text{dB}\mu\text{V}) - 3(\text{dB}). \tag{9.14}$$

If, for example, $V_{3\text{IP}} = 82$ dBμV the level of the blocking signal will be $\tfrac{3}{2} \times 82 - 3 = 120$ dBμV or 1 V. Because of the high levels involved it is quite possible that the fifth-order intermodulation products may also be of significant amplitude; if so, $V_B = \frac{5}{4}V_{5IP} - 4.5 \text{ dB}\mu\text{V}$.

Sensitivity

The sensitivity of a communication radio receiver is the smallest input signal voltage that is required to give a specified output power with a specified output signal-to-noise ratio. It is necessary to include signal-to-noise ratio in the definition because otherwise the output could consist mainly of noise and be of little use. The lower limit to the sensitivity of a receiver is set by the input thermal noise plus some contribution from the noise factor of the receiver.

For an amplitude-modulated receiver typical figures might be:

(a) 2 μV sensitivity with 30% modulation at 400 Hz and 20 dB signal-to-noise ratio (h.f. receiver);

(b) 1 μV sensitivity with 50% modulation at 1 kHz for 12 dB SINAD (p. 203) and 7.5 kHz selectivity (v.h.f. receiver); and

(c) −99 dBm sensitivity with 30% modulation at 1 kHz for 12 dB SINAD and 8 kHz selectivity (h.f. receiver).

An s.s.b. receiver will have a better sensitivity than a d.s.b. receiver because there is no carrier power and the bandwidth is narrower. Typically, sensitivity = 0.5 μV with 1 kHz output for 12 dB SINAD and 3 kHz selectivity.

For a frequency-modulation receiver the sensitivity is quoted with a specified r.m.s. or peak frequency deviation. Typically, this might be sensitivity 0.5 μV with 2.1 kHz r.m.s. frequency deviation (this is 3 kHz peak deviation) for 12 dB SINAD and 15 kHz selectivity.

If a receiver employs one, or more, tuned r.f. stages its sensitivity will vary with frequency as shown by Example 9.4.

Example 9.4

A radio receiver can be tuned to receive signals in the frequency band 4 to 20 MHz with the tracking error given by Table 9.1. At 20 MHz the sensitivity of the receiver is 2 μV. Calculate its sensitivity at (a) 4 MHz, (b) 8 MHz, and (c) 14 MHz and plot the sensitivity curve of the receiver. Assume the r.f. stage to have a Q-factor of 50.

Table 9.1

Signal frequency (MHz)	4	8	14	20
Tracking error (kHz)	20	80	100	0

Solution
When the receiver is tuned to any particular frequency any tracking error appears in the r.f. stage.

(*a*) When the wanted signal is at 4 MHz the r.f. stage is tuned to 4.02 MHz. Hence

$$\text{sensitivity} = 2 \times R_d \bigg/ \left[\frac{R_d}{\sqrt{(1 + Q^2 B^2 / f_0^2)}} \right] \mu V$$

$$= 2 \sqrt{\left[1 + \frac{50^2 (40 \times 10^3)^2}{(4.02 \times 10^6)^2} \right]} \mu V = 2.23 \ \mu V.$$

(*Ans.*)

(*b*) The r.f. stage is tuned to 8.08 MHz. Hence

$$\text{sensitivity} = 2 \sqrt{\left[1 + \frac{50^2 (160 \times 10^3)^2}{(8.08 \times 10^6)^2} \right]} = 2.81 \ \mu V.$$

(*Ans.*)

(*c*) The r.f. stage is tuned to 14.1 MHz. Hence

$$\text{sensitivity} = 2 \sqrt{\left[1 + \frac{50^2 (200 \times 10^3)^2}{(14.1 \times 10^6)^2} \right]} = 2.45 \ \mu V.$$

(*Ans.*)

The sensitivity curve of the receiver is shown plotted in Fig. 9.5.

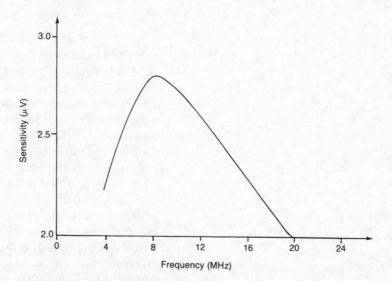

Fig. 9.5 Sensitivity curve of a radio receiver.

Selectivity

The selectivity of a communication radio receiver is its ability to select a wanted signal from all the signals that are simultaneously present at the aerial terminals and to reject all other signals. Selectivity may be quoted graphically, showing the output of the receiver in decibels relative to the maximum output, plotted against frequency off-tune. Alternatively, several points on such a graph may be quoted; e.g. 6 dB down at 3 kHz bandwidth, 60 dB down at 12 kHz bandwidth. Most of the selectivity of a radio receiver is provided by the i.f. filters. The 6 dB and 60 dB bandwidths are known, respectively, as the *nose* and the *skirt bandwidths*. The nose bandwidth is the band of frequencies over which a signal can be received with little loss of strength.

The skirt bandwidth is the band of frequencies over which it is possible to receive a strong signal. The *shape factor* is the ratio (skirt bandwidth)/(nose bandwidth) and typically it is about 4.

The *adjacent-channel ratio* of a receiver is the ratio, in decibels, of the input voltages at the wanted, and at the adjacent-channel frequencies, necessary for the adjacent channel to produce an output power of 30 dB down on the wanted signal power.

In modern h.f. communication receivers the selectivity is provided mainly by crystal filters, whereas v.h.f./u.h.f. receivers often employ SAW filters. With the use of either of these types of filter it is possible to achieve almost any desired selectivity.

Noise Factor

The output of a radio receiver must always contain some noise, partly because the input signal is not noise free and partly because the receiver itself generates some noise. The noise factor F of a radio receiver is a measure of the degradation of the input signal-to-noise ratio caused by the receiver, i.e.

$$F = \frac{\text{input signal-to-noise ratio}}{\text{output signal-to-noise ratio}}. \tag{9.15}$$

At all frequencies up to about 30 MHz the noise picked up by an aerial is generally larger than the noise internally generated by the receiver. There is then little to be gained by the receiver having a low noise factor and often an r.f. amplifier is not provided. At higher frequencies the aerial noise is much smaller than the internally generated noise and then a low noise factor is advantageous. Consequently, v.h.f./u.h.f. receivers always employ r.f. gain.

The noise factor of an m.f./h.f. radio receiver is typically in the region of 10 to 12 dB; v.h.f./u.h.f. receivers have a typical noise factor of 8 to 10 dB but, in some cases, it may be only about 3 dB.

Since the sensitivity of a radio receiver is defined in terms of a specified output signal-to-noise ratio it is evident that sensitivity and noise factor are related. If the sensitivity of a receiver is N μV for an output signal-to-noise ratio of S (as a ratio) in a bandwidth of B Hz, then the noise factor of the receiver is

$$F = 61 + 20 \log_{10} N - 10 \log_{10} (S - 1) - 10 \log_{10} B \text{ dB}. \tag{9.16}$$

Example 9.5

A radio receiver has a sensitivity of 1.5 μV for an output signal-to-noise ratio of 20 dB in a bandwidth of 3 kHz. Calculate its noise factor.

Solution
From equation (9.16)

$$F = 61 + 20 \log_{10} 1.5 - 10 \log_{10} 99 - 10 \log_{10} 3000 = 9.8 \text{ dB}.$$
(*Ans.*)

SINAD Ratio

Sometimes it is more helpful to consider the total distortion at the output of a receiver as well as the noise. The SINAD ratio is given by equation (9.17), i.e.

$$\text{SINAD} = \frac{(\text{signal power} + \text{noise power} + \text{distortion power})}{(\text{noise power} + \text{distortion power})}. \qquad (9.17)$$

The sensitivity of a radio receiver is often specified in terms of SINAD.

Stages in a Radio Receiver

The Radio-frequency Stage

The radio-frequency stage, or *front end*, of a communication radio receiver has several functions to perform.

(*a*) It must couple the aerial to the receiver in an efficient manner. A wide variety of coupling circuits can be used and Fig. 9.6 shows just a few of them. The tuning capacitances will probably be provided by voltage-tuned varactor diodes.

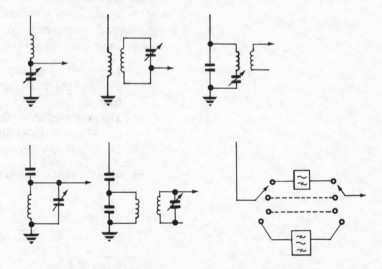

Fig. 9.6 Some r.f. coupling circuits.

(*b*) It must suppress signals at the image-channel and intermediate frequencies. For less stringent requirements tuned circuits will be able to provide sufficient selectivity, but for optimum performance a bank of sub-octave filters will be necessary. The bank of filters will divide up the tuning range of the receiver into sub-bands and any one of the filters is switched into circuit at a given time. Some h.f. receivers which employ up-conversion may use only a single 30 MHz low-pass filter. In a v.h.f./u.h.f. receiver the band-pass filters may be helical resonators. When filters are used to obtain the r.f. selectivity any r.f. amplifiers employed will be wideband circuits.

(c) Older-type communication receivers tended to employ an r.f. amplifier at frequencies in excess of about 5 MHz. The modern tendency is *not* to employ r.f. gain below about 30 MHz, because little, if any, improvement in the noise performance of the receiver results, and an r.f. amplifier is a source of both cross modulation and intermodulation.

When an r.f. amplifier is employed it must be designed to have a low noise factor, to operate linearly for even the strongest anticipated input signals, and to generate the minimum intermodulation products. If the amplifier is a broadband type it will be susceptible to second-order intermodulation products as well as to third-order ones. The gain of the r.f. amplifier may be varied by the a.g.c. system of the receiver, or the gain may be constant and the amplifier preceded by an a.g.c.-controlled r.f. attenuator.

The Mixer Stage

The function of the first mixer stage is to convert the wanted signal frequency into the first intermediate frequency. Similarly, the function of the second mixer is to convert the first intermediate frequency into the second intermediate frequency. Double-balanced mixers are increasingly used for several reasons.

(a) They give a high degree of isolation between the first local oscillator and the r.f. stage which minimizes unwanted radiation from the aerial.

(b) The local oscillator voltage fed into the i.f. filter is suppressed by at least 30 to 40 dB and this reduces noise.

(c) The even-order intermodulation performance is good.

Probably the most commonly employed discrete-component double-

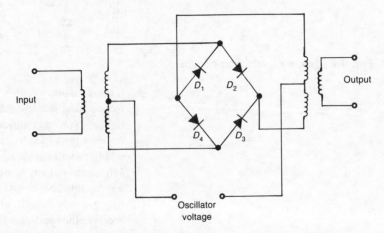

Fig. 9.7 Diode ring mixer.

balanced mixer is the diode ring, shown in Fig. 9.7, largely because it has a large dynamic range. Ring mixers are commercially available as a complete package ready for use. The circuit does have a number of disadvantages; these are: (*i*) the local oscillator power must be fairly high; (*ii*) the circuit has a conversion loss of at least 6 dB; (*iii*) the intermodulation performance depends upon the source and load impedances; and (*iv*) the rejection of the carrier component depends upon the balance of the circuit. Most of the difficulties can be overcome with the use of a *transistor-tree* balanced mixer whose basic circuit is shown in Fig. 9.8. This circuit is employed in several ICs and an example is the Plessey SL 6440.

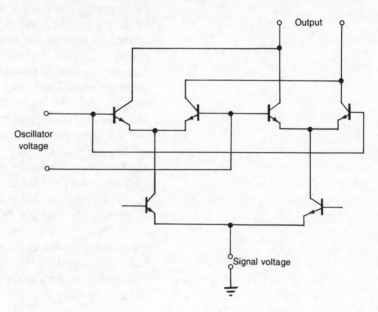

Fig. 9.8 Transistor-tree balanced mixer.

The Local Oscillator

The first local oscillator must be capable of tuning to any frequency in the tuning range of the receiver *plus* the first intermediate frequency. The second local oscillator has to provide only one frequency, this is equal to the sum of the first and the second intermediate frequencies.

The first local oscillator must have: (*a*) high *spectral purity*; (*b*) *frequency agility*, so that it can quickly change frequency; (*c*) small *increments of frequency* (generally, for the frequency bands below 30 MHz a frequency resolution of 1 to 100 Hz is required, with s.s.b. receivers no worse than 10 Hz; v.h.f./u.h.f. receivers usually have a quoted frequency resolution of about 1 kHz); and (*d*) *frequency stability*.

The above requirements are difficult to satisfy with an *LC* oscillator and if the receiver is to operate at a few fixed frequencies a crystal oscillator with switched crystals may be used. Increasingly, modern communication receivers employ a *frequency synthesizer* to generate the two local oscillator frequencies. A frequency synthesizer is able to generate a large number of precise frequencies which are derived from a single, stable, reference source. The disadvantage that is introduced by the use of a frequency synthesizer is that the tuning of the receiver is not continuous but can only occur in discrete steps.

Frequency Synthesizers

A frequency synthesizer is a circuit that derives a large number of discrete frequencies, singly or simultaneously, from an accurate, high-stability crystal oscillator source. Each of the derived frequencies has the accuracy and the stability of the reference source. A frequency synthesizer must be able to cover a wide frequency band so that the receiver can work over the whole of the tuning range. Most of the frequency synthesizers employed in modern communication receivers are of the *phase-locked loop* (p.l.l.) type and Fig. 9.9 illustrates the basic concept. The system consists of a very stable crystal oscillator which acts as the reference source, a phase detector, a low-pass filter, and a voltage-controlled oscillator (v.c.o.). The phase detector produces an output d.c. control voltage, the magnitude and polarity of which is determined by the phase difference between the crystal oscillator and v.c.o. voltages. The control voltage is filtered to remove all a.c. components and it is then applied to the v.c.o. to vary its frequency. The action of the p.l.l. ensures that the frequency of the v.c.o. changes in the direction that reduces any difference between the crystal oscillator frequency and the v.c.o. frequency. Once *lock* has been achieved the two inputs to the phase detector are at the same frequency, but there is always a phase difference between them in order to maintain the controlling d.c. voltage.

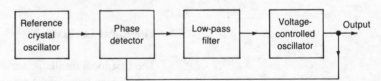

Fig. 9.9 Phase-locked loop.

To obtain more than one output frequency a frequency divider, which may be a programmable type, must be connected in the position shown in Fig. 9.10(*a*). The signals applied to the phase detector are then at frequencies of f_R and f_C/N_1, and the v.c.o. runs at a frequency of $f_C = N_1 f_R$. The p.l.l. with a frequency divider in the loop allows a large number of frequencies to be obtained by altering the division ratio N_1. Each of the possible output frequencies is an integral multiple of the reference frequency. If, for example, $f_R = 1$ MHz and $N_1 = 3$ the v.c.o. frequency will be 3 MHz, but if $N_1 = 20$ the frequency of the v.c.o. will be 20 MHz. Clearly, the increments in frequency that can be obtained are equal to the reference frequency

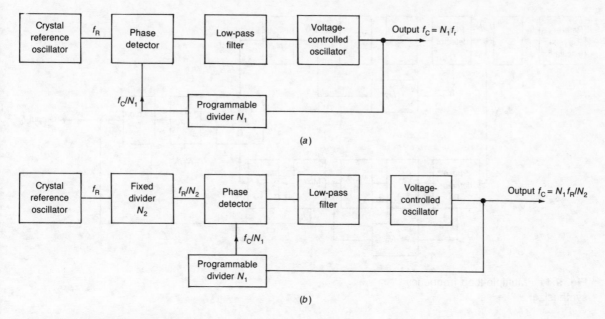

Fig. 9.10 Frequency synthesizers: (a) using a p.l.l., and (b) with improved frequency resolution.

f_R. To improve the frequency resolution the reference frequency can also be divided, as shown by Fig. 9.10(b). If, now, $f_R = 1$ MHz and $N_2 = 100$ the output frequency will be $f_C = N_1 f_R / N_2$ with a frequency resolution of 1 kHz. The method is relatively simple, fully digital, and can be integrated, but it is slow to change from one frequency to another.

To obtain both rapid frequency changes and small frequency resolution a multiple-loop frequency synthesizer is often employed, and Fig. 9.11 shows one example of the technique. The top loop produces an output at frequency $f_R N_A / N_1$ and this is divided down by the ratio N_2 to give frequency $f_R N_A / N_1 N_2$. The lower loop produces an output at frequency $f_R N_B / N_1$ and this is applied to a mixer, along with the output of the system at frequency f_0. The band-pass filter selects the difference frequency $f_0 - f_R N_B / N_1$ so that the inputs to the phase detector C are at frequencies $f_R N_A / N_1 N_2$ and $f_0 - f_R N_B / N_1$; these are locked by the output loop to become equal to one another. Then

$$f_0 = \frac{f_R}{N_1} \left[\frac{N_A}{N_2} + N_B \right]. \tag{9.18}$$

Example 9.6

In the frequency synthesizer of Fig. 9.11 $f_R = 1$ MHz, $N_1 = 10$ and $N_2 = 100$. Determine the range of output frequencies of the synthesizer if N_A is variable from 200 to 300 and N_B from 350 to 400.

Solution
From equation (9.18)

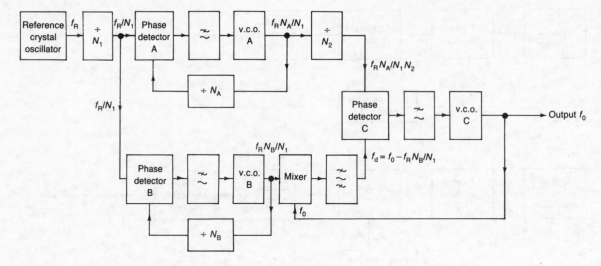

Fig. 9.11 Multiple-loop frequency synthesizer.

$$f_{0(min)} = \frac{1 \times 10^6}{10} \left[\frac{200}{100} + 350 \right] = 35.2 \text{ MHz}$$

$$f_{0(max)} = \frac{1 \times 10^6}{10} \left[\frac{300}{100} + 400 \right] = 40.3 \text{ MHz}.$$

Therefore, the range is from 35.2 MHz to 40.3 MHz.   (*Ans.*)

Ganging and Tracking

Older types of communication receiver employ mechanical control of the tuning of the receiver. The rotation of a tuning-control knob simultaneously varies tuning capacitors in both the r.f. stage and the first local oscillator. The tuning capacitors are mounted on a common spindle and are said to be *ganged*. The maintenance of the correct frequency difference (the first intermediate frequency) between the frequencies to which the r.f. stage and the oscillator are tuned is called the *tracking*. Usually, identical capacitors are used with different values of inductance in each circuit. Tracking errors are inevitable and they result in a variation of both the sensitivity (see Example 9.4) and the image-channel rejection of the receiver. More modern radio receivers replace the tuning capacitors with voltage-tuned varactor diodes; this allows rapid frequency changing to take place, often under microprocessor control. Modern communication receivers often employ a frequency synthesizer and up-conversion and are then able to avoid tracking problems by using either no r.f. selectivity at all or a bank of switched band-pass filters.

Intermediate-Frequency Amplifier Stages

The function of the i.f. amplifier in a communication receiver is to provide most of the gain and the selectivity of the receiver. Each i.f. amplifier stage must shape and select a relatively narrow bandwidth at the mixer output and reject adjacent-channel signals. The second i.f. stage must amplify the signal to the level necessary for the detector to operate satisfactorily. In the past, double-tuned coupled circuits were commonly employed to provide a desired loss–frequency characteristic but, since they have a poor shape factor, they are rarely used today. Most modern communication receivers employ either *crystal filters* or *SAW filters*, to give the desired selectivity. These filters offer the considerable advantages of requiring no i.f. amplifier alignment, and having a selectivity that is not affected by the application of a.g.c. to the stage.

Crystal Filters

Standard crystal filters (see *Radio Systems for Technicians*) are readily available at a number of fixed frequencies, e.g. 100 kHz, 1.4 MHz, 10.7 MHz and 35.4 MHz, with a bandwidth of between 0.01% to 1% of the centre frequency. The insertion loss of a crystal filter is between 1 dB and 10 dB, the shape factor is very good and the generation of spurious responses is small.

SAW Filters

A *surface acoustic wave* (SAW) filter is a four-terminal structure that has a pair of comb-like transducers deposited onto a piezo-electric substrate (see Fig. 9.12). The input transducer converts an electrical signal into a surface acoustic wave, while the output transducer converts the acoustic signal back to electrical form. The 'fingers' of each transducer are spaced apart by a common distance d. When an input voltage is applied to the SAW filter a surface acoustic wave is excited that propagates along the substrate. The maximum excitation is obtained when the comb spacing is equal to the wavelength λ of the signal. Half of the acoustic power is transmitted to the absorber and is completely lost, the other half is sent in the opposite direction towards the output transducer. This means that the minimum loss of

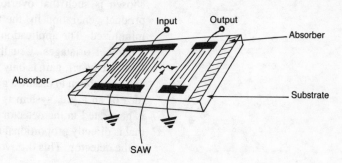

Fig. 9.12 SAW filter.

the SAW filter is 6 dB. When the surface acoustic wave passes through the output transducer it is converted into an electrical signal.

When the frequency of the signal is not at the nominal centre frequency, the comb spacing is no longer equal to the acoustic signal wavelength and the electric/acoustic conversion efficiency is reduced. The roll-off of the conversion efficiency is very rapid and so a highly selective loss–frequency characteristic is obtained. The characteristic can be tailored by the manufacturer by varying the length and/or the number of the comb teeth and/or their spacings in each transducer. Like crystal filters SAW filters are offered by the manufacturers at a number of set frequencies, e.g. 45 MHz, 100 MHz and 405 MHz, with a bandwidth of between 1% and 10% of the centre frequency. Their insertion loss varies from 6 dB to about 28 dB. The SAW filter is frequently used to provide i.f. selectivity because it is small, lightweight, very reliable and it requires no adjustments.

The Detector Stage

The function of the detector stage is to recover the information modulated on to the received carrier, and often also to generate the a.g.c. voltage. Most a.m. receivers that employ discrete circuitry in the detector stage still use the diode detector because of its simplicity and its good performance. The main problem is that the input signal level must be several times larger than the threshold level of the diode, otherwise considerable signal distortion will occur. The demodulation of an s.s.b. or c.w. signal requires the use of a product detector. Very often the detection process is carried out within an IC that also provides a number of other circuit functions. Frequency-modulation receivers tend to use the ratio detector in discrete-component designs, and either the quadrature detector or the phase-locked loop detector if ICs are employed.

Automatic Gain Control

The amplitude of the wanted carrier that appears at the input of a radio receiver may fluctuate widely, by perhaps 100 dB or more. *Automatic gain control* (a.g.c.) is applied to a receiver to maintain the carrier level at the detector input at a more or less constant value. The level chosen is such that overload of, and consequent intermodulation product generation in, the final r.f. stage and/or the first mixer is minimized. The application of the a.g.c. voltage is distributed over a number of stages. Usually, the gain of the i.f. stages is reduced first and the r.f. gain is only reduced when the level of the input signal is large enough to ensure a good output signal-to-noise ratio. The basic idea of an a.g.c. system is illustrated by Fig. 9.13. A d.c. voltage is generated in the detector stage (or in a separate a.g.c. generator) that is directly proportional to the amplitude of the carrier at the input to the detector. This d.c. voltage is applied to each of the controlled

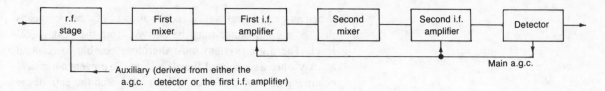

Fig. 9.13 Application of a.g.c. to a radio receiver.

stages to vary their voltage gains. If the carrier level should rise the a.g.c. voltage will also increase and will reduce the gain of each controlled stage. This will, of course, reduce the overall gain of the receiver and so tend to restore the carrier level at the detector input to its original value. Conversely, if the carrier level should fall the a.g.c. system will increase the overall gain of the receiver.

Whenever the r.f. input signal is large and is likely to cause overloading of one, or more, stages the auxiliary a.g.c. will come into action and reduce the gain of the r.f. stage. If the a.g.c. voltage is used to vary the gain of an r.f. amplifier problems may arise with regard to both its dynamic range and the production of distortion because of shifts in the operating point of the amplifier(s). An alternative, that overcomes these problems, is the use of an a.g.c.-controlled r.f. attenuator. The r.f. attenuator may either have a continuous loss that is varied by the a.g.c. voltage, or have fixed values of loss that are switched into, or out of, circuit by the a.g.c. voltage. The r.f. attenuators are often fitted in front of, and in between, the stages of r.f. gain as shown by Fig. 9.14.

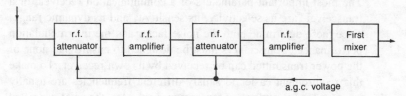

Fig. 9.14 Use of r.f. attenuators to vary the gain of a receiver.

All a.g.c. systems have an inherent delay in their response to a change in the input signal voltage. It is not desirable for the a.g.c. system to have too rapid a response or it will respond to impulsive noise. The *attack time* is the time taken for the a.g.c. voltage to rise to a predetermined percentage of its final value when the carrier level falls. The *decay time* is the time taken for the a.g.c. voltage to fall to a predetermined percentage of its original value when the carrier level rises. The choice of the attack and delay times depend upon the kind of signal being received.

(a) For a d.s.b.a.m. signal the constant-frequency carrier can be used to generate the a.g.c. voltage. The attack and delay times need only be fast enough to allow the a.g.c. system to respond to fading, but slow enough to avoid it responding to low-frequency modulation. Typical figures are in the region 0.1 s to 0.2 s.

(*b*) When an s.s.b. signal is received the absence of a carrier means that the a.g.c. voltage must be derived from the peak signal level. The a.g.c. system must therefore be able to respond quickly when a modulated signal appears. To prevent a transient occurring at the end of each syllable of speech the gain of the receiver must only slowly increase at the end of each syllable. If the attack time is too long the system may not be able to follow rapid fades but, on the other hand, if the attack time is too short each new syllable will be accompanied by a 'roaring' sound. Typically, the attack time should be from 2 to 10 ms and the decay time 500 ms to 1 s.

There are three ways in which the a.g.c. system for an s.s.b. receiver can be improved. These are: (*i*) the use of a pilot carrier (which would also be useful for automatic frequency control), (*ii*) the use of *hang a.g.c.*, in which envelope-derived a.g.c. is sustained for about 0.6 s after the signal has fallen to zero, and (*iii*) a combination of envelope a.g.c. with a fast-acting squelch circuit that operates during the inter-syllable intervals.

The performance of an a.g.c. system is quoted in manufacturer's literature in the form, 'less than 6 dB change in output voltage for 90 dB increase from threshold'. Many frequency-modulated receivers are not provided with an a.g.c. system but instead rely solely upon limiting to keep the detected output signal at a constant level.

Communication Receivers

The most important parameters of a communication receiver, or a transceiver, are its selectivity, its sensitivity and its dynamic range, since these determine both the noise factor and the intermodulation performance. Transceivers must be designed to ensure that none of the power transmitted can be received by its own receiver. To make this requirement easier to satisfy different frequencies are usually employed for reception and for transmission. Also, the noise generated within the transmitter should not increase the level of the received noise. Low-power transmitters often have several stages of r.f. filtering before the r.f. power amplifier output stage to reduce off-tune noise.

Figure 9.15 shows the block diagram of the signal circuitry of the Eddystone 1650 m.f./h.f. communication radio receiver, which operates over the frequency range of 10 kHz to 30 MHz in 5 Hz steps. The receiver can operate with d.s.b. amplitude-modulated, s.s.b. (both lower- and upper-sideband), and c.w. signals. Six different second i.f. 6 dB bandwidths are provided, namely 300 Hz, 1 kHz, 2.4 kHz, 3 kHz, 8 kHz and 16 kHz, by switching into circuit the appropriate band-pass filter. The r.f. a.g.c. voltage is generated by an a.g.c. generator IC and this voltage is used to vary the loss of the pin diode r.f. attenuator. When the product detector is switched into circuit an audio a.g.c. generator is used to produce the a.g.c. voltage for the i.f. amplifier.

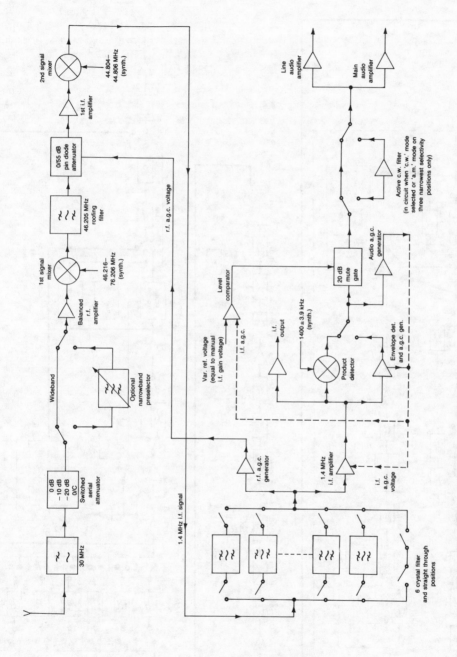

Fig. 9.15 The Eddystone 1650 m.f./ h.f. communication radio receiver. (*Courtesy of Eddystone Radio Ltd.*)

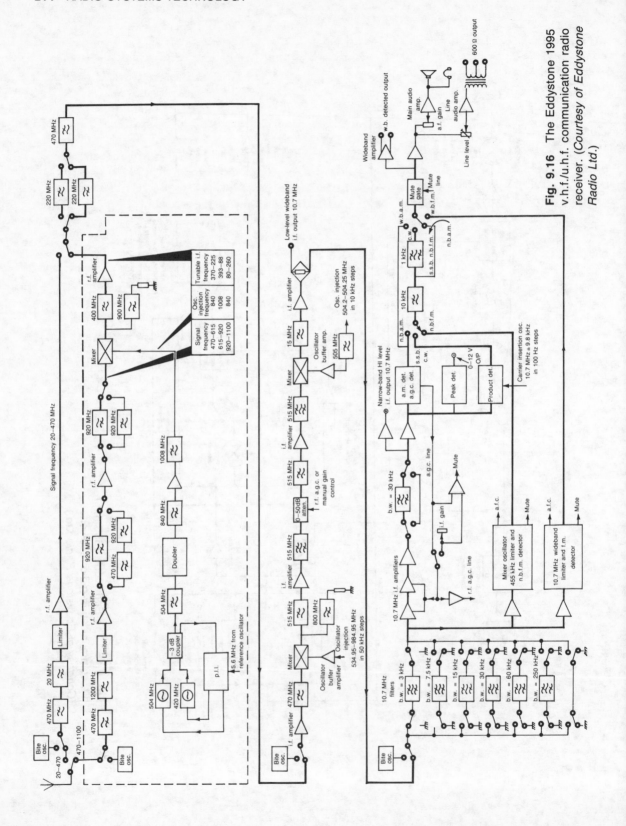

Fig. 9.16 The Eddystone 1995 v.h.f./u.h.f. communication radio receiver. *(Courtesy of Eddystone Radio Ltd.)*

The block diagram of the Eddystone 1995 v.h.f./u.h.f. communication radio receiver is shown by Fig. 9.16. The blocks marked as 'bite' are built-in test equipment oscillators that are provided to aid fault location. The receiver covers the frequency band 20 MHz to 1.1 GHz; for frequencies between 470 MHz and 1.1 GHz a front-end down-converter is used to shift the signal to the frequency band 80 to 393 MHz. A cheaper variant of the receiver covers only the frequency band 20 MHz to 470 MHz and this does not need the down-converter. The main part of the receiver has a first i.f. of 515 MHz with a roofing bandwidth (maximum overall bandwidth) of 6 kHz, and a second i.f. of 10.7 MHz. The bandwidth of the second i.f. filter can be set in the range 3 kHz, 7.5 kHz, 15 kHz, 30 kHz or 60 kHz by choosing one of the five crystal filters, to 250 Hz using a ceramic filter, or to 600 Hz using a roofing filter. The band-limited output of the second i.f. filter is then applied to the appropriate detector for the type of signal being received; this may be a.m., or f.m. narrowband or wide-band. For amplitude modulation detection may be achieved by an envelope, or a product detector; n.b.f.m. detection is carried out at 455 kHz and wideband f.m. detection at 10.7 MHz. Audio muting, or squelch, is provided to reduce noise whilst tuning the receiver from one signal to another. Complete muting is not used so that low-level signals are not missed. The muting is signal level-derived for all reception modes other than f.m. when it is noise derived.

Both the Eddystone 1650 and the 1995 series of communication radio receivers are operated with microprocessor control and they provide scanning, sweeping and channel-storage facilities. Each receiver can be human operated, or computer operated, or operated from a distance by a remote-control unit.

10 Radio Systems

The public telecommunications network of a country is used for the transmission of speech, telegraphy, data and sound/television broadcast signals. All the circuits employed, except for those in the local distribution network, are routed over multi-channel telephony systems or high-bit-rate data systems which, in turn, are routed via some combination of copper cable, optical fibre, or microwave radio-relay system. Line-of-sight radio-relay systems are extensively used to provide a wide range of communication services. Their traffic capacity varies from a few, to several thousand, speech channels using either analogue or digital techniques. Besides the public network there are also a number of private telecommunication networks operated by such organizations as the railways, gas and electricity companies.

International telecommunication networks involve the use of both copper and optical-fibre cable, both underground and submarine, terrestrial radio-relay and communications satellite systems. The traditional application of satellite communications has been the international telecommunications network but this is increasingly being challenged by optical-fibre systems. Communications satellite systems cannot compete effectively with optical fibre but they will continue to carry a large proportion of the total long-distance traffic for a long time to come. New satellite applications, including television broadcasting and point-to-point business services, such as British Telecom's SATSTREAM, are emerging all the time.

Land-mobile systems are commonly employed both for private networks, such as ambulance, police and taxis, and for communication via the public switched telephone network (p.s.t.n.). Comprehensive mobile systems also exist for both air and maritime communications.

Microwave Radio-relay Systems

A microwave radio-relay system employs line-of-sight space-wave transmissions in both the u.h.f. and the s.h.f. bands. Nearly always, the length of a route is much longer than the maximum possible distance between two aerials and then a number of radio-relay stations, or repeaters, are necessary. The basic idea of a microwave radio-relay system is illustrated by Fig. 10.1. The radio signal radiated by the transmitting aerial is received by the first relay station where it is amplified before it is re-transmitted to the next relay station, and so on until the signal arrives at the receiving station. The r.f. signal must be amplified before its amplitude has fallen to such a level that the minimum required signal-to-noise ratio cannot be obtained. This

Fig. 10.1 Microwave radio-relay system.

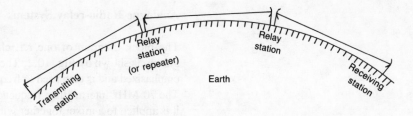

factor determines the spacing between, and hence the number of, the relay stations.

Most of the relay systems presently in use operate in frequency bands below 10 GHz and employ analogue techniques with frequency modulation. Analogue systems suffer from a progressive deterioration in the signal-to-noise ratio with increase in the length of the route and this limits the performance of a system. The more modern radio-relay systems employ digital techniques using either quadrature amplitude modulation or some form of phase-shift modulation. Since a digital signal can be regenerated at each relay station a required signal-to-noise ratio can be maintained throughout the length of a system.

Table 10.1 lists the frequency bands that are in use for microwave radio-relay systems in the UK.

Table 10.1

Frequency (GHz)	Name	Use
1.7–1.9	2 GHz spur	Television and 960 channel analogue. Spurs and low-capacity links.
1.9–2.3	2 GHz main	960 channel analogue (baseband 316–4188 kHz).
3.7–4.2	4 GHz band	1800 channel analogue (baseband 316–8204 kHz). 140 Mb/s digital. 2 × 34 Mb/s digital.
5.85–6.425	Lower 6 GHz band	
6.425–7.11	Upper 6 GHz band	960 channel analogue. 140 Mb/s digital. 2 × 34 Mb/s digital. Television.
10.7–11.7	11 GHz band	
14.0–14.5	14 GHz band	
17.7–19.7	19 GHz band	8 Mb/s digital medium-capacity feeders. 140 Mb/s digital.

Notes:
A 140 Mb/s system gives 11 520 64 kb/s channels.
The 2 × 34 Mb/s systems give 720 64 kb/s channels.
The 2 × 34 Mb/s systems are used for short-spur and junction connections.

Analogue Radio-relay Systems

The block diagram of one r.f. channel in an analogue radio-relay system is shown by Fig. 10.2. The input baseband signal is first pre-emphasized and is then used to frequency modulate a 70 MHz carrier. The 70 MHz intermediate frequency is then amplitude limited before it is applied to a mixer together with the output of a microwave oscillator. This oscillator runs at a frequency equal to $(f_t - 70)$ MHz, where f_t is the frequency at which the signal is to be radiated from the aerial. The upper sideband $[(f_t - 70) + 70]$ MHz $= f_t$, is selected by the sideband filter and is then amplified by either a travelling-wave amplifier (t.w.a.) or a solid-state amplifier. The amplified signal then passes through first an isolator and then another filter, before it is routed to the transmit aerial via one, or more, circulators. The circulators allow the odd-numbered r.f. channels to be combined together and radiated as a horizontally polarized wave. Similarly, the even-numbered r.f. channels are also combined together and are radiated with vertical polarization. At the receiver the signal is selected by the appropriate band-pass channel filter and is then mixed with the output of a microwave oscillator. This oscillator runs at a frequency f_0 of $(f_t + 70)$ MHz and so the mixer output contains a component at $(f_t + 70) - f_t = 70$ MHz; this component is selected by the i.f. filters. The 70 MHz signal is first amplified and equalized

Fig. 10.2 One r.f. channel in an analogue radio-relay system. (From *British Telecommunication Engineering*.)

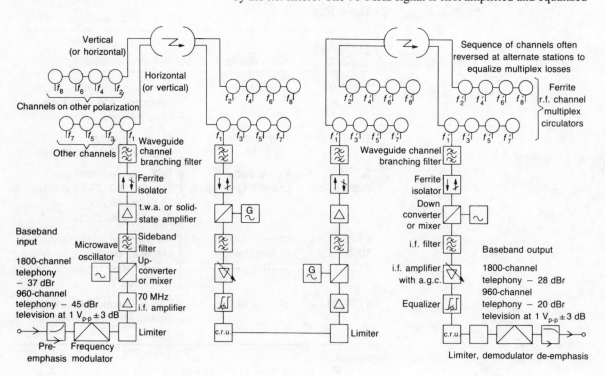

TRANSMIT TERMINAL   REPEATER   RECEIVE TERMINAL

before it is demodulated to recover the baseband signal. Finally, the baseband signal is de-emphasized to obtain the original amplitude relationships between the low-, and the high-frequency components.

The block marked as c.r.u. is the carrier re-insertion unit. Its function is to inject a noise-free signal into the signal path whenever the incoming carrier fades completely. This prevents the a.g.c. system of the receiver increasing the gain and producing a large output noise. The development of solid-state Gunn oscillators and their use with microwave cavities has led to the disappearance in radio-relay equipment of many low-power microwave thermionic devices. The travelling wave tube (t.w.t.) or amplifier (t.w.a.), which can provide a high-power output with a relatively high efficiency is, however, still employed in both analogue and digital systems. The power output that semiconductor amplifiers can develop at gigahertz frequencies is steadily increasing and such circuits are increasingly employed as the r.f. power amplifier. At present, a solid-state amplifier can only handle some tens of watts compared with a t.w.a.'s hundreds of watts. In modern equipment all the other electronic circuitry uses semiconductor devices. Microwave Ga or InP ICs have been developed which can work up to about 30 GHz in microstrip circuitry.

Digital Radio-relay Systems

The main disadvantage of an analogue radio-relay system is that the noise powers generated in each link are additive and so the signal-to-noise ratio inevitably decreases with increase in the length of a route. The problem can be overcome by the use of a digital system, since a digital signal can be regenerated at each relay station and then the noise is no longer cumulative. Further advantages arising from the use of digital techniques are as follows.

(a) The ever-increasing availability of l.s.i. and v.l.s.i. digital ICs allows cheaper, smaller-sized, equipment to be designed and used.

(b) The widespread use of ICs reduces power consumption and increases reliability.

(c) Converting a system from analogue to digital operation can considerably increase its channel capacity.

(d) The bit error rate for data signals sent over a digital system is much lower than if the same signals are transmitted over an analogue system.

Digital radio-relay systems operate in a number of frequency bands (see Table 10.1) with bit rates of either 2×34 Mb/s or 140 Mb/s. The former can give 720, and the latter 11 520, 64 kb/s telephone channels. The 140 Mb/s rate has been standardized by the CCIR because it is compatible with the fourth hierarchical digital multiplexing level of 139.264 Mb/s. In the UK the standard digital radio-relay systems operate in the 11 GHz band and they provide six 140 Mb/s r.f. channels. This gives a capacity of 6 × 1920, or 11 520, telephone channels.

Digital Modulation

Digital modulation of a carrier may modulate either the amplitude, the frequency, or the phase of that carrier, but usually some form of *phase-shift keying* is employed. With bi-phase p.s.k., or b.p.s.k., the modulating signal puts the carrier phase into either one of two possible states. With quaternary p.s.k., or q.p.s.k., the signal is coded into the dibits 00, 01, 11 and 10 and these are each represented by one of the four possible phases of the carrier. A combination of amplitude and phase modulation, known as quadrature amplitude modulation (q.a.m.) is also used in some systems.

With b.p.s.k. the carrier is transmitted with the reference phase, i.e. 0°, to indicate binary 1, and with the opposite phase to indicate binary 0. Differential b.p.s.k. means that the reference phase is the phase of the last bit, or dibit, received; it has the advantage that the receiver does not need an absolute phase reference. With q.p.s.k. phase shifts of +45°, +135°, +225° and +315° relative to the phase of the previous symbol represent, respectively, the dibits 00, 01, 11 and 10. This is shown by what is known as a *constellation* (see Fig. 10.3(a)). A q.a.m. signal has 2^n phases, each of which can have more than one amplitude. Figure 10.3(b) gives the constellation of a q.a.m. signal with eight different phases and two different amplitudes; this gives 16 different states and it will allow combinations of four bits, e.g. 0101 or 1010, to be coded.

Figures 10.4(a) and (b) show, respectively, the block diagrams of the transmitter and the receiver of an 11 GHz, 140 Mb/s, digital radio-relay system. As with an analogue system, a relay station, or repeater, consists of the back-to-back linking of a receiver and a transmitter, and the necessary links are shown by the dashed lines.

The input signal is modulated by a pseudo-random digital sequence that scrambles its frequency spectrum. The scrambling process is used

Fig. 10.3 Constellations of (a) q.p.s.k. and (b) q.a.m. signals.

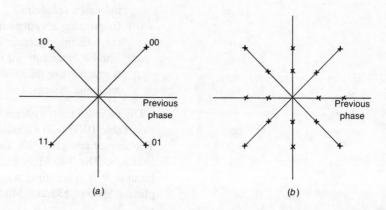

(a)

(b)

because: (*a*) it improves the spectral and power distributions of the signal; (*b*) it reduces jitter;† (*c*) it makes the system appear transparent; and (*d*) it provides timing information. The scrambled signal is split into two 70 Mb/s streams and each of these is applied to a four-phase modulator. Also applied to the phase modulator is an 11 GHz carrier signal produced by the frequency multiplication of the output of a 1 GHz oscillator. The output of the modulator is a four-phase p.s.k. signal with a symbol rate of 70 Mbaud. This signal is then both amplified and filtered before it is combined with the other r.f. channels and fed to the transmit aerial. The final stage of amplification may be provided with either a t.w.a. or a solid-state circuit.

At the receiver the incoming 11 GHz signal is selected by the appropriate channel filter and it is then applied to a balanced mixer. Here it is mixed with an 11 GHz signal to produce upper and lower sidebands. The lower sideband, centred on 140 MHz, is selected by the i.f. filter; it is then amplified, before it is applied to the phase-locked loop demodulator. The reference carrier for demodulation is obtained by multiplying the received i.f. signal by 4; this process removes the phase modulation and gives a reference carrier at $4f_{IF}$.

Multiplexing the outputs of the six r.f. channels is carried out using a combination of circulators and horizontal/vertical polarization. This is shown by Fig. 10.5.

18 GHz equipment operating at 2 Mb/s and 8 Mb/s is also employed and is often used to provide Megastream circuits.

Fig. 10.4 The 11 GHz, 140 Mb/s digital radio-relay system, (*a*) transmitter and (*b*) receiver. (From *British Telecommunication Engineering*.)

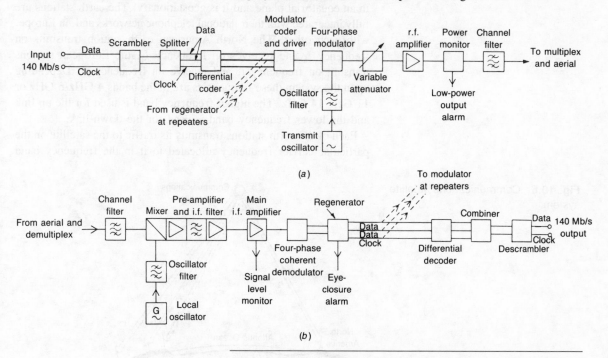

(*a*)

(*b*)

†Jitter is the short-term variation of the significant instants of a digital signal from their ideal position in time. Effectively it is phase modulation of the signal timing.

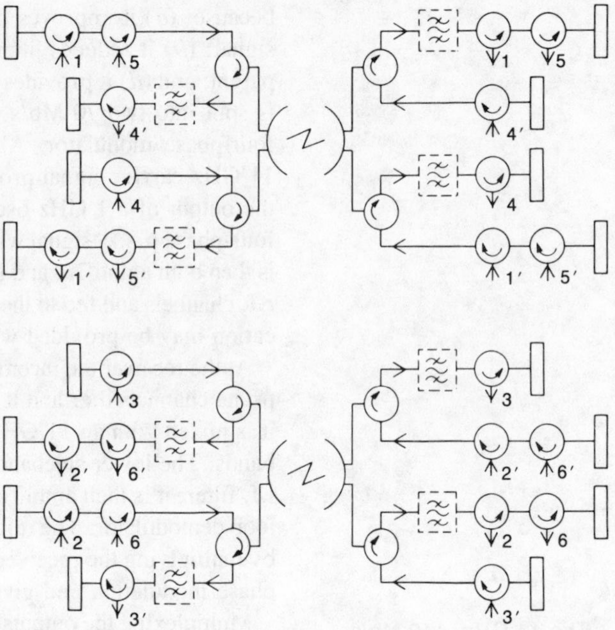

Fig. 10.5 Multiplexing r.f. channels using circulators. (From *British Telecommunication Engineering*.)

Communications Satellite Systems

The basic principle of a communications satellite system is illustrated by Fig. 10.6. The satellite is in an orbit 35 800 km above the earth in an equatorial plane and it is geostationary. The earth stations are fully integrated with their national telephone networks and, in Europe, with each other. The North American earth station transmits on frequency f_1 and receives on frequency f_4, the European stations transmit on frequency f_3 and receive on frequency f_2. Depending upon the system these frequencies are in the bands 4 GHz/6 GHz or 11 GHz/14 GHz. The higher frequency band is used for the up-link and the lower frequency band is used for the down-link.

Each of the earth stations transmits its traffic to the satellite on the particular carrier frequency allocated to it in the frequency band

Fig. 10.6 Communications satellite system.

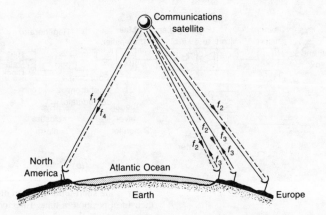

5.935−6.425 GHz, or 14−14.5 GHz. This is a bandwidth of 500 MHz and it allows the simultaneous use of a satellite by more than one earth station. The capacity of a satellite is shared between its various users using either *frequency-division multiple access* (f.d.m.a.) or *time-division multiple access* (t.d.m.a.). The number of telephone channels thus provided to a user varies from 12 in a 2.5 MHz bandwidth to 1872 in a bandwidth of 36 MHz. All the signals transmitted to earth by a satellite are received by every earth station; each station selects the particular carrier frequencies that it has been allocated, in the frequency bands of either 3.7−4.2 GHz or 10.95−11.2 GHz and 11.45−11.7 GHz.

Communications satellites that form an integral part of the international telephone network are operated by the *Communication Satellite Corporation* (COMSAT) on behalf of an international body known as INTELSAT (International Telecommunication Satellite Consortium). The capacity of an INTELSAT satellite has increased considerably since the first satellite was put into service in 1965. This is shown by Table 10.2.

The Western European countries have launched their own satellite, known as EUTELSAT, which is employed to carry relatively local traffic.

Figure 10.7 shows the block diagram of the communication equipment of an INTELSAT V satellite. The satellite has seven receiving aerials: east and west hemi, east and west zone, and global at 6 GHz, and east and west spot at 14 GHz. The signals received by each aerial are amplified and are then applied to a mixer to be shifted to the common 4 GHz band. The amplified mixer output is then applied to various band-pass filters to obtain the required transponder band-

Table 10.2

INTELSAT No.	Date of first launch	Bandwidth (MHz)	Capacity
I	1965	50	240 channel or TV
II	1967	130	
III	1968	300	1500 channel or TV
IV	1971	500	4000 channel or TV
IVA	1975	800	6000 channel or TV
V	1980	2144	12 000 channel or TV
VA	1985	2250	15 000 channel or TV
VI	1987	3300	30 000 channel or TV

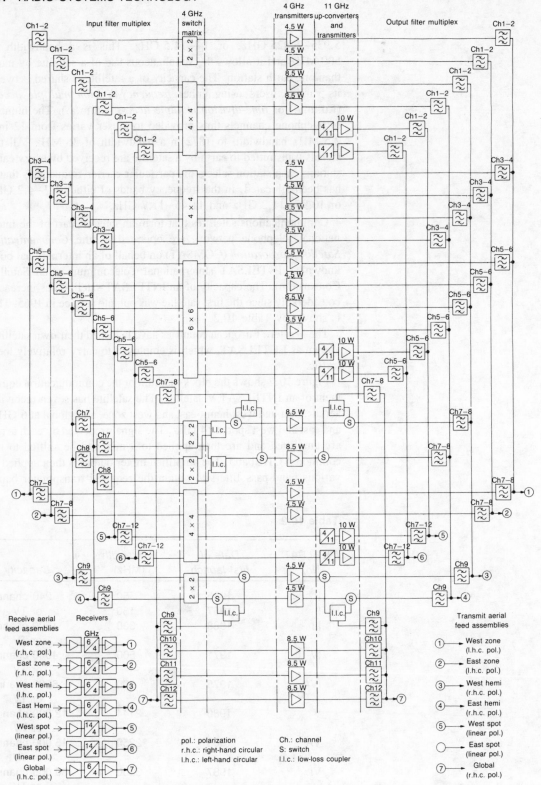

Fig. 10.7 INTELSAT V satellite: communication circuitry. (From *British Telecommunication Engineering*.)

widths. (Transponder is a term used in satellite technology to denote a wideband channel.) The band-limited signals are then directed to a switch matrix which is controlled from the ground. Most of the outputs of the switch matrix are amplified by a t.w.a. Some switch matrix outputs, however, are first frequency shifted to the 11 GHz band and then they are amplified, again by a t.w.a. The t.w.a.s have saturated output powers of either 4 W or 8.5 W at 4 GHz and of 10 W at 11 GHz, but they are normally operated at a lower power level in order to reduce intermodulation. The output of each t.w.a. is then applied to the output filter multiplex and hence to one of the seven transmit aerials.

Multiple Access

Multiple access is the name given to the way in which the traffic-carrying capacity of a satellite can be considerably increased. It allows a large number of earth stations to be given access to a satellite and it may employ either frequency-division, or time-division techniques. The number of telephone channels made available to an earth station may be fixed (this is known as *fixed access*) or the allocated number may be assigned in response to the traffic demand (this is known as *demand access*).

Frequency-division Multiple Access

In frequency-division multiple access (f.d.m.a.) all the earth stations which share a communication satellite do so at the same time, each station being allocated its own unique frequency bands. The f.d.m.a. system is always employed in conjunction with both frequency modulation and frequency-division multiplexing. Each earth station is allocated one, or more, carrier frequency(ies) and it modulates all of its out-going traffic, regardless of its destination, on to that carrier. Every earth station in the network must therefore be able to receive at least one carrier from all the other earth stations.

Large blocks of telephone channels are allocated to individual earth stations on a semi-permanent basis. The number of circuits required on each route can be calculated from traffic studies. For some routes which carry little traffic only a few circuits are required and this results in their having a poor availability. On some other routes there may not be enough traffic to economically justify the provision of even one circuit. In any case, because the telephone traffic varies with the time of day it is unlikely that all the allocated channels would be in use for all of the time. This means that the fixed access version of f.d.m.a. is not efficient. On the routes with lower traffic density the transponder's bandwidth can be divided into a large number of carriers and a particular channel allocated to each one. This system is known as *single-channel per carrier* (s.p.s.c.) working.

If demand access is employed a pool of channels is made available to all the earth stations. A channel is only assigned to a particular route between two earth stations as the demand arises. The receiving station is notified which channel is to be used before transmission starts, and when the connection is no longer required the channel is returned to general availability status. The control of a demand-access f.d.m.a. system is exercised by a control channel and computers.

The number of demand-assignment telephone channels needed to carry a certain telephone traffic can be further reduced by the use of *speech interpolation*. A digital speech interpolation (DSI) equipment monitors the telephone conversations and fills any gaps in the speech with speech from other calls. The activity is monitored by the DSI equipment and every time a speaker pauses the equipment may take away the channel and assign it to another call. A speaker will not be disconnected unless the channel is wanted for another active call. When the speaker speaks again a new channel is assigned to him. Since, on average, more than 50% of a conversation consists of silent intervals, the circuit capacity can be at least doubled.

SPADE

A digital system, known as *single-channel per carrier p.c.m. multiple-access demand-assignment equipment* (SPADE) is a demand-assignment system which uses a separate r.f. carrier for each telephony channel. The bandwidth of the satellite's transponder is divided into 800 channels which are associated in pairs to give 400 two-way circuits. Control of the setting-up of a connection between two earth stations is vested in the *demand assignment and switching unit* (d.a.s.s.) that is provided at each station. The necessary communication links between d.a.s.s. units is provided by a common signalling channel which is shared between all the earth stations on a time-division basis.

When a call request is received the d.a.s.s. unit selects a pair of carrier frequencies and informs the destination earth station that an incoming call is imminent and which frequencies are to be used. The analogue signal is then applied to a p.c.m. encoder to produce a 56 kb/s p.c.m. signal. This signal is then placed into a memory and is read out at the higher rate of 64 kb/s. This leaves some gaps in the signal which are filled, using a process known as *bit stuffing*, with preamble and start-of-message synchronization bits. The 64 kb/s signal is then applied to a four-bit p.s.k. modulator to produce a q.p.s.k. signal. This q.p.s.k. signal is processed by the earth station's transmitting equipment and is radiated by the aerial.

Time-division Multiple Access

Time-division multiple access (t.d.m.a.) is a technique which allows a number of earth stations to have access to a common satellite trans-

Fig. 10.8 Time-division multiple access.

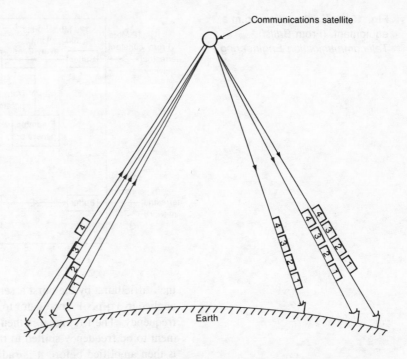

ponder at different times. Each earth station takes its turn to transmit data through the transponder for a small fraction of the total time, as shown by Fig. 10.8. The *bursts* of data from each of the *n* users arrive at the satellite in a pre-assigned sequence. The bandwidth occupied by a burst depends upon the type of modulation used and it may, or may not, occupy the whole of the bandwidth of the transponder. As long as each earth station is able to maintain the correct instants for its bursts to start and to finish its bursts will not overlap the bursts generated by other earth stations. In the INTELSAT systems the p.c.m. digital signals that are transmitted have an 8 kHz sampling rate so that the pulses are 125 μs wide. The frame size is 2 ms and hence each traffic burst includes $(2 \times 10^{-3})/(125 \times 10^{-6}) = 16$ samples from each of the telephone channels being transmitted.

The basic block diagram of the equipment needed at each t.d.m.a. earth station is shown by Fig. 10.9. The input digital data is fed into a buffer store at 12 Mb/s and is read out of the store at the much higher rate of 120 Mb/s. Since the output bit rate is much higher than the input bit rate it is clear that the data can only be transmitted in short bursts. The bursts are repeated every 2 ms and for the same input and output information $12 \times 10^6 = (120 \times 10^6 T)/(2 \times 10^{-3})$, and so the time duration T of a burst is 200 μs. Before it is transmitted each traffic burst must have a *preamble* added; the preamble is the name given to a number of bits which are added to the traffic burst to enable a distant earth station to correctly receive the signal. Preamble bits are generated in the t.d.m.a. equipment and added to

Fig. 10.9 Earth-station t.d.m.a. equipment. (From *British Telecommunication Engineering*.)

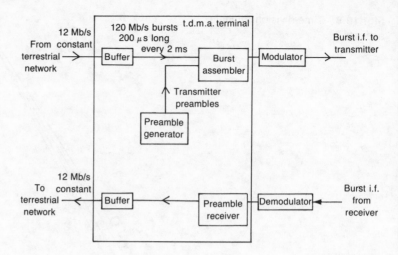

the traffic burst by the burst assembler. The assembled burst is then applied to a q.p.s.k. modulator to produce a burst at the intermediate frequency. The i.f. signal is then applied to the transmitting equipment to be frequency shifted to the allocated frequency band and it is then amplified before it is radiated from the aerial. A receiving earth station must demodulate the received carrier and then recover the synchronization pulses before it will be able to identify the beginning of each frame. It will then be able to assemble the original telephony signal and pass it on to the destination.

A t.d.m.a. system must be synchronized to make sure that the traffic bursts originating from different earth stations do not overlap one another. The necessary synchronization is obtained by the use of *reference bursts*. The reference bursts are transmitted by some of the earth stations and they are received by all of the earth stations in the network. Each earth station must then ensure that its traffic bursts, with added preamble, are transmitted to the satellite at fixed times relative to each reference burst.

Radiation Techniques

The traffic capacity of a communication satellite can still further be increased by the use of (*a*) spatial polarization, and (*b*) polarization diversity. Spatial polarization means that separate transponders are employed for (*i*) wide area coverage, and (*ii*) for localized spot coverage, of the earth. Polarization diversity means that the same frequency band can be used without mutual interference by two separate signals if one signal is transmitted using horizontal polarization and the other signal uses vertical polarization. Alternatively, left- and right-hand circular polarizations are sometimes used.

Communication Satellite Orbits

Most communications satellites are in orbit at a height of 35 800 km above the earth in an equatorial plane. At this height the orbiting satellite keeps pace with the rotation of the earth and so it appears to be stationary above a particular point on the earth's surface. This orbit is therefore known as the *geo-synchronous orbit*. Satellites in the geo-synchronous orbit must be sufficiently far apart for interference from up-links to adjacent (in the orbit) satellites not to be a problem. This means that the earth stations must use highly directive aerials, and the CCIR have specified the maximum sidelobe level which is allowable. The specified aerial gain G is

$$G = 29 - 25 \log_{10} \theta \text{ dBi}, \tag{10.1}$$

where θ is the angular distance in degrees between satellites in the geo-synchronous orbit. Each degree corresponds to a span of about 700 km in that orbit.

Example 10.1

An earth station requires an e.r.p. of 80 dBW in order to obtain the required signal-to-noise ratio at the satellite. Two parabolic dish aerials are available; one aerial has a gain of 40 dBi and, hence, a 40 dBW signal, the other aerial has a 60 dBi gain and a 20 dBW signal. Determine which aerial should be used if $\theta = 2°$.

Solution

Since both of the aerials are able to supply the wanted e.r.p. the choice of aerial is based upon the sidelobe levels of their radiation patterns. The first aerial will produce an interference e.r.p. of

$$40 + 29 - 25 \log_{10} 2 = 61.5 \text{ dBW}.$$

The second aerial produces an interference e.r.p. of

$$20 + 29 - 25 \log_{10} 2 = 41.5 \text{ dBW}.$$

Assuming the adjacent satellite also receives a wanted 80 dBW signal the interference is 18.5 dB down for the first aerial and 38.5 dB down for the second aerial. Therefore the second aerial should be chosen.   (*Ans.*)

Noise Performance of Analogue Microwave Links

An analogue microwave system, be it terrestrial or satellite, carries a number of 3.1 kHz bandwidth telephony channels that have been assembled, using frequency-division multiplex, to form a baseband signal. The number of telephony channels that are *active*, i.e. carrying speech, at any instant, and hence the mean baseband power, varies continuously and reaches its maximum value in the busy hour. (Note that on some routes the 'busy hour' may be as much as 2 or even 2.5 hours long.) If the system is heavily loaded, i.e. most of its channels are carrying speech, there will be an increase in the level of intermodulation noise. On the other hand, if the system is lightly loaded,

Fig. 10.10 Showing how the noise-to-carrier ratio varies with the frequency deviation.

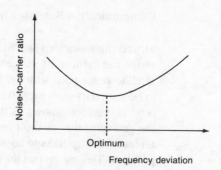

the mean baseband power will be low and, since the thermal noise is at a constant level, the carrier-to-noise ratio will fall. This means that the carrier-to-noise ratio increases with increase in the baseband signal level, and hence with the number of active channels, up to an optimum level above which increased intermodulation noise ensures that the carrier-to-noise ratio falls. In turn, this means that there is an optimum value for the frequency deviation of the microwave carrier in the transmitter. This is shown by Fig. 10.10. The CCIR state that the mean power of a single telephone channel in a system having 240 channels, or more, is -15 dBm0† or 32 μW. The CCIR also give the mean power of N multiplexed channels as

$$P = -15 + 10 \log_{10} N \text{ dBm0}, \tag{10.2}$$

where $N > 240$.

If there are fewer than 240 channels then the N channels have a mean power of

$$P = -1 + 4 \log_{10} N \text{ dBm0}. \tag{10.3}$$

A microwave link is set up so that a sinusoidal 1 kHz, 0 dBm0 test tone produces a standard value of frequency deviation. This standard frequency deviation is 200 kHz for a 960 channel system and 140 kHz for a 1800 channel system. The relationship between the test-tone frequency deviation and the frequency deviation produced by the baseband signal must be known in order to ensure that the i.f. bandwidth is wide enough. The i.f. bandwidth must be able to accommodate the i.f. signal produced by the *peak* level of the baseband signal.

The signal-to-noise ratio at the output of a telephone channel is the ratio of the level of the standard test tone (0 dBm0 at 1 kHz) to the noise in the 3.1 kHz bandwidth. The worst-case telephone channel is the top channel, i.e. the one with the highest f.d.m. carrier frequency. The r.m.s. test-tone frequency deviation $f_{td(r.m.s.)}$ is related to the r.m.s. baseband frequency deviation $f_{d(r.m.s.)}$ by the *loading factor L*, which is given by equation (10.2) or by equation (10.3).

---

†dBm0 = dBm − dBr. If a signal level is measured at any point in a system it can be related to the zero transmission level point by the relative level in dBr of that point.

The loading factor is also equal to

$$L = 20 \log_{10}\left(\frac{\text{r.m.s. baseband deviation}}{\text{r.m.s. test-tone deviation}}\right)$$

$$= 20 \log_{10} l. \tag{10.4}$$

The ratio of the *peak* frequency deviation produced by the baseband signal to the *r.m.s.* frequency deviation is known as the *peak factor p*. For systems with more than about 25 channels the peak factor is approximately equal to 3.2. The peak baseband frequency deviation $f_{d(peak)}$ is equal to $pf_{d(r.m.s.)} = lpf_{td(r.m.s.)}$, and hence the bandwidth occupied by the microwave signal is given by equation

$$B = 2(pf_{d(r.m.s.)} + f_{max}). \tag{10.5}$$

Example 10.2

A microwave carrier is frequency modulated by the baseband signal representing 1800, 3.1 kHz bandwidth, telephony channels arranged in frequency-division multiplex. The top channel is at 8204 kHz and the r.m.s. test-tone frequency deviation is 140 kHz. Calculate (a) the r.m.s. frequency deviation produced by the baseband signal, and (b) the minimum r.f. channel bandwidth necessary. Assume the peak factor to be 3.16.

Solution

(a) From equation (10.2)

$$L = -15 + 10 \log_{10} 1800 = 17.6 \text{ dBm0}.$$

Therefore,

$$17.6 = 20 \log_{10}\left(\frac{f_{d(r.m.s.)}}{140 \times 10^3}\right)$$

and

$$f_{d(r.m.s.)} = 7.59 \times 140 \times 10^3 = 1.063 \text{ MHz}. \quad (Ans.)$$

(b) From equation (10.5)

$$\text{Minimum bandwidth} = 2(3.16 \times 1.063 \times 10^6 + 8.202 \times 10^6)$$
$$\simeq 23.1 \text{ MHz}. \quad (Ans.)$$

Test-tone Analysis

When a frequency-modulated carrier of peak voltage V_c has a single-frequency noise voltage V_n superimposed upon it the carrier will be both frequency, and phase modulated (p. 36). The peak frequency deviation of the carrier is given by equation (2.23) as $V_n f_{diff}/V_c$, where f_{diff} is the separation between the carrier frequency and the interfering signal frequency. In a 1 Ω resistance $V_n = \sqrt{(kT)}$ and the carrier power is $P = V_c^2/2$. Provided the carrier-to-noise ratio is at least 10 dB, the peak frequency deviation of the carrier is

$\sqrt{(kT/P_c)}f_{diff}$. Hence, the r.m.s. frequency deviation is given by

$$f_{td(r.m.s.)} = \sqrt{\left(\frac{kT}{2P_c}\right)}f_{diff}. \tag{10.6}$$

Because of the triangular noise spectrum (p. 38) the worst-case channel is the channel that is located at the highest carrier frequency f_{max} in the multiplexed baseband signal. If the bandwidth of each telephony channel is b Hz then the noise power N_0 at the output of the top channel is given by

$$N_0 = 2K \int_{f_{max}-b}^{f_{max}} \left(\frac{kT}{2P_c}f_{diff}^2\right) df. \tag{10.7}$$

The factor 2 is necessary because the input noise is at frequencies both above and below the carrier frequency, and K is the transfer constant of the frequency detector.

$$N_0 = \frac{KkT}{3P_c}\left[f_{max}^3 - (f_{max} - b)^3\right] = \frac{KkTf_{max}^2 b}{P_c}$$

since

$$(f_{max} - b)^3 \simeq f_{max}^3\left(1 - \frac{3b}{f_{max}}\right) = f_{max}^3 - 3f_{max}^2 b.$$

The output signal power is proportional to the square of the r.m.s. frequency deviation of the carrier, i.e. $S_0 = Kf_{td(r.m.s.)}^2$, and so the output signal-to-noise ratio is

$$\frac{f_{td(r.m.s.)}^2 P_c}{kTf_{max}^2 b} = \frac{f_{td(r.m.s.)}^2 B_{IF} P_c}{kTB_{IF}f_{max}^2 b}$$

or   output signal-to-noise ratio $= \left[\left(\frac{f_{td(r.m.s.)}}{f_{max}}\right)^2 \frac{B_{IF}}{b}\right]$

times the carrier-to-noise ratio. $\tag{10.8}$

Usually, the output signal-to-noise ratio is quoted in decibels. Thus

output signal-to-noise ratio = input carrier-to-noise ratio (dB)

$$+ 20 \log_{10}\left(\frac{f_{td(r.m.s.)}}{f_{max}}\right) + 10 \log_{10}\left(\frac{B_{IF}}{b}\right) \text{dB}. \tag{10.9}$$

Sometimes the frequency deviation of the carrier due to the test tone is smaller than the maximum baseband frequency and then the f.m. advantage will be negative. The last term is always positive and it expresses the fact that a single telephony channel does not occupy all of the i.f. bandwidth.

The output signal-to-noise ratio is increased by the use of pre-emphasis in the transmitter and de-emphasis in the receiver. Figure 10.11 shows the pre-emphasis characteristic that is recommended by the CCIR; it gives a signal-to-noise ratio improvement of 4 dB.

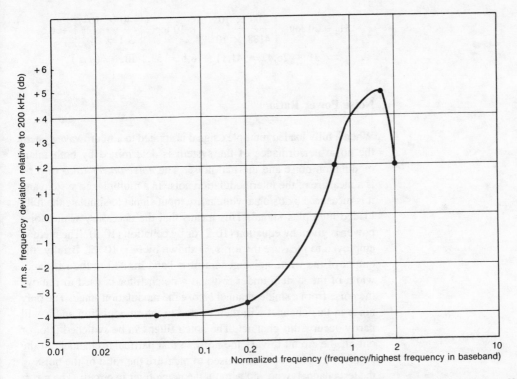

Fig. 10.11 CCIR pre-emphasis characteristic.

Furthermore, the use of psophometric weighting effectively gives another 2.5 dB increase in the output signal-to-noise ratio.

Example 10.3

The carrier power at the input to a radio-relay receiver is −87 dBW. The receiver has an i.f. bandwidth of 40 MHz, a noise factor of 10 dB, and the carrier frequency deviation due to a 1 kHz 0 dBm0 test tone is 200 kHz. The carrier is frequency modulated by a baseband signal consisting of the frequency-division multiplex of 960 3.1 kHz bandwidth telephony channels. Calculate the test-tone signal-to-noise ratio at the output of the top telephone channel at 4188 kHz.

Solution

$$F = 10 \text{ dB} = 10; \ P_c = -87 \text{ dBW} = 2 \text{ nW}.$$

The input noise power is

$$kT_0B = 1.38 \times 10^{-23} \times 290 \times 40 \times 10^6 = 160 \times 10^{-15} \text{ W}.$$

The noise at the output of the detector is

$$FGkT_0B = 160 \times 10^{-14}G.$$

Therefore, the carrier-to-noise ratio is

$$\frac{2 \times 10^{-9}G}{160 \times 10^{-14}G} = 1249.4 = 31 \text{ dB}.$$

From equation (10.9), the output signal-to-noise ratio is

$$31 + 20 \log_{10}\left[\frac{200 \times 10^3}{4188 \times 10^3}\right] + 10 \log_{10}\left[\frac{40 \times 10^6}{3.1 \times 10^3}\right] + 6.5$$

$$= 31 - 26.42 + 41.11 + 6.5 = 52.2 \text{ dB.} \quad (Ans.)$$

Noise Power Ratio

When a fully loaded multiplex signal is applied to a microwave system the noise performance of the system is determined by both inter-modulation noise and thermal noise. The *noise power ratio* (n.p.r.) is a measure of the intermodulation noise in a multiplexed system and it is measured by using a white noise input signal to simulate the fully loaded multiplex signal. This means that the necessary white noise power is given by equation (10.2) or by equation (10.3). The method employed to measure the n.p.r. is shown by Fig. 10.12. Briefly, the output of the white noise generator is band limited to the i.f. bandwidth of the system under test and a notch filter is used to remove the noise from a single channel *before* the modulation stage. The only noise in this channel will then be due to intermodulation and it will partly occupy this channel. The notch filter can be switched into, or out of, the circuit to vary the noise in that particular channel. At the receiver a *noise receiver* is used to measure the ratio of the noise in the test channel, with and without the notch filter in circuit. The n.p.r. is then the difference between the two power levels, quoted in decibels. Thus, referring to Fig. 10.12, the n.p.r. is equal to $10 \log_{10} P_1/P_2$ dB. Typically, the n.p.r. of a microwave system is in the region of 50 dB.

A measured value of n.p.r. can be converted into a corresponding value of signal-to-noise ratio by the use of equation (10.10) or equation (10.11).

$$\text{Signal-to-noise ratio} = \text{n.p.r.} + 61 \log_{10} N + 4.6 \text{ dB} \tag{10.10}$$

for $N < 240$ channels.

$$\text{Signal-to-noise ratio} = \text{n.p.r.} + 18.8 \text{ dB} \tag{10.11}$$

for $N > 240$ channels.

Fig. 10.12 Measurement of noise power ratio.

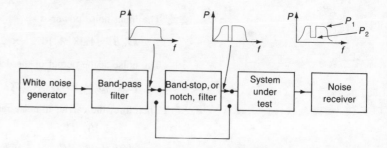

Example 10.4

Calculate the n.p.r. of a 960 channel system that has a worst-case channel signal-to-noise ratio of 50 dB.

Solution
From equation (10.11),

$$50 = \text{n.p.r.} + 18.8 \text{ dB}$$

or   n.p.r. = 31.2 dB.   *(Ans.)*

Bit Error Rate in Digital Systems

The performance of a digital microwave system, be it terrestrial or satellite, is expressed in terms of its *bit error rate* (b.e.r.). The b.e.r. is equivalent to the output signal-to-noise ratio of an analogue system and it is the probability that a transmitted bit will be incorrectly received. The b.e.r. is quoted as a number, e.g. 1×10^{-5} means that, on average, one bit in every 100 000 will be in error.

An error in the correct reception of a bit may occur because noise picked up by the system has corrupted the signal waveform to such an extent that the decision circuitry in the receiver cannot accurately determine whether a bit is a 1 or a 0. The CCITT recommendations for a digital radio-relay link are: (*a*) a b.e.r. of 1×10^{-6} over a one-minute period must not be exceeded for more than 0.4% of any month; and (*b*) a b.e.r. of 1×10^{-3} over a one-second period must not be exceeded for more than 0.054% of any month.

Expressions are available which relate b.e.r. to signal-to-noise ratio for the various forms of digital modulation but they are beyond the scope of this book.

Land-mobile Radio Systems

Many organizations, both private and public, depend upon mobile radio systems for their successful operation. Examples are many and include: (*a*) *emergency* (the ambulance, fire and police services); (*b*) *public utilities* (gas, water and electricity); and (*c*) *private* (delivery vans, service technicians, mini-cabs and taxis). Despite the inclusion of emergency services and the public utilities these land-mobile services are normally lumped together and referred to as the *private land-mobile radio* (PMR) service. This is to distinguish services that do *not* have access to the *public switched telephone network* (p.s.t.n.) from the *public* land-mobile services which do have such access. Nowadays, this latter service is provided by *cellular radio*, and it now provides national coverage of the UK. *Paging systems* are also commonly employed, particularly in hospitals, to alert a person that he is required to contact his office.

Private Mobile Radio

The channel spacing for a PMR system has been standardized at 12.5 kHz, and several users may have to share a channel in overlapping service areas. Both amplitude and frequency modulation are used in conjunction with *double-frequency simplex* operation. This term means that two frequencies are used for each call, one for each direction of transmission but that only one person may speak at a time. Since frequencies in either the v.h.f. or the u.h.f. band are employed, the area covered by a base station is limited by both the e.r.p. and the height of its aerial. The coverage area is made as large as possible to reduce costs to a minimum, but often a number of interconnected base stations are necessary to cover a large area. Figure 10.13 shows a typical arrangement. The mobiles are not given any particular frequencies but are allocated a pair of frequencies on demand.

Fig. 10.13 Land-mobile radio system.

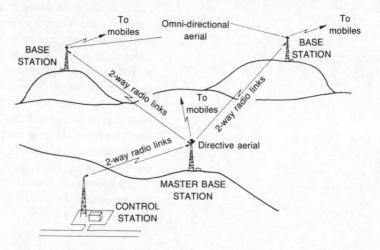

In the older PMR systems a car, or van, driver is required to keep his receiver turned on and he must listen for his call-sign to be broadcast. The driver can then respond to the call-sign and receive his call. Modern PMR systems employ *five-tone selective calling*. An individual mobile can be contacted by means of its identifying code which is a unique combination of five tones. The mobile receiver recognizes its code and automatically responds to the call. With so many organizations, big and small, joining the ranks of land-mobile users the demand for frequencies has become intense. One solution to the problem of the limited available frequency spectrum is the use of *trunking*. Trunking is the name given to the use of a number of channels that are shared between a larger number of mobiles. When a mobile wishes to make a call the radio searches the allocated frequency band for a free channel, and when one is found it is seized and used to contact the base station. Once a channel has been allocated to a particular mobile that allocation lasts for the duration of the call. Immediately the call is terminated the channel is released and is returned to the 'pool' for re-use. A trunked system can handle up to six times as much traffic as a basic PMR system whilst appearing to the users as though each had the exclusive use of a channel.

Cellular Radio

A *cellular radio* system divides up the geographic area to be covered into a large number of much smaller areas or cells. Each cell has its own low-power base station sited somewhere within the cell. The cells are grouped together in clusters, as shown by Fig. 10.14, and each cluster has the available radio channels allocated to it in a regular pattern that is repeated over the entire area. Each base station is allocated a particular set of channels for telephonic communication. Each set is a proportion of the total number of channels allocated to the system; the remaining channels are allocated to other cells in the cluster. This process allows the same channel frequencies to be re-used many times within the other clusters. The number of cells in a cluster must be such that the clusters can fit together into contiguous areas; in practice, this means 4-cell, 7-cell, 12-cell and 21-cell clusters. Figure 10.15(*a*) shows how three 7-cell clusters fit together; adjacent cells in different clusters clearly do *not* share the same frequencies. The frequency allocation for each cluster is shown in Fig. 10.15(*b*); transmissions from base to mobile are in the frequency band 890 MHz to 935 MHz and in the reverse direction they are in the band 935 MHz to 980 MHz. When the number of cells in a cluster is small there will be more channels allocated to each cell, and this means that a cell can carry a greater telephone traffic. Unfortunately, reducing the number of cells in each cluster also reduces the distance between two cells using the same channel frequencies, and this increases co-channel interference. The 7-cell cluster is probably the best compromise between these two conflicting factors and it is for this reason that the 7-cell cluster is the most often employed. Not all the channels are

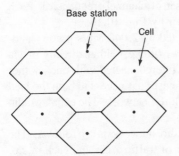

Fig. 10.14 A cell cluster in a cellular radio system.

Fig. 10.15 (*a*) Three 7-cell clusters. (*b*) Frequency allocation for each cluster in (*a*).

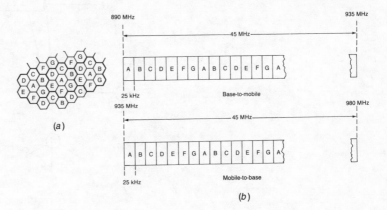

used for telephone conversations: 21 channels are used for the provision of common signalling facilities.

The geographic size of a cell is varied in accordance with the anticipated telephone traffic. In a rural area a cell may have a diameter of 30 km or more, but in an inner-city area the cell diameter may be less than 2 km. Reducing the size of a cell provides more cells, and hence more channels, in an area of a given size but, since the cell separation is then reduced, it also increases co-channel interference. Co-channel interference can be reduced by the use of *sectored aerials* at the base station. A three-sectored aerial has a coverage angle of 120°, a six-sectored aerial has a coverage of 60°, compared with the 360° coverage provided by an omni-directional aerial. A cell is then divided into three, or six sectors, each of which effectively becomes a new cell with its own set of channel frequencies. Each cell is now corner-excited as shown by Fig. 10.16.

The base stations are interconnected by 2.048 Mb/s p.c.m. landline links to *mobile switching centres* (m.s.c.) either directly or via a nodal point. The m.s.c. are fully interconnected with one another and with the p.s.t.n. so that a mobile user has access to both the land public telephone network and the cellular network. The cellular radio network is illustrated by Fig. 10.17.

The operation of the cellular radio network is as follows. The network is organized into a number of traffic areas and each m.s.c. keeps track of the location of each mobile. Whenever a mobile is not active it continuously receives from the nearest m.s.c. a code that identifies the traffic area in which it is travelling. If the received code

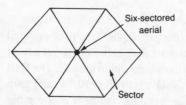

Fig. 10.16 Use of a six-sectored aerial.

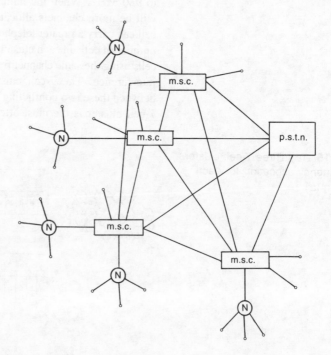

Fig. 10.17 Cellular radio network.

is not error free, indicating a low signal level, the mobile will search for another signal with a higher signal level. Once the new signal has been found the mobile will check whether the traffic area code has changed. If it has, it means that the mobile has moved into another traffic area and the mobile will then register its new location by identifying itself to the new base station. The network then ensures that all the m.s.c. have recorded the new location of the mobile.

To originate a call a mobile transmits a message over a common signalling channel to the nearest base station and it is then allocated a channel. If, during the duration of a call, the mobile moves from one cell to another the call is automatically taken over by the new base station. A base station continuously monitors the signal level received from all the mobiles in its cell and if this level falls below a threshold value it informs the m.s.c. The m.s.c. then commands all the surrounding base stations to measure the signal level that they are receiving from the mobile. The m.s.c. will then transfer the call to the base station that is receiving the highest signal level from the mobile, and it notifies both base stations of the new channel frequency. The original base station then sends a control signal to the mobile which switches its transceiver to the new channel. The changeover, known as *hand-off*, takes place so quickly that the user is only aware of a very brief break in transmission.

Table 10.3

Frequency band (MHz)	Maximum number of channels	Channel spacing (kHz)	Peak frequency deviation (kHz)	Signalling bit rate (kb/s)
890–935 (base to mobile) 935–980 (mobile to base)	1800	25	9.5	8

The cellular radio system used in the UK is known as the *Total Access Communication System* (TACS) and it has the parameters shown in Table 10.3.

Exercises

Chapter 1

1.1 A carrier wave has an r.m.s. value of 4 A when unmodulated which rises to 4.4 A when it is sinusoidally modulated. Calculate the depth of modulation. The carrier is then modulated by a bipolar square wave having the same peak value as the sine wave. Determine the r.m.s. value of the modulated wave.

1.2 A carrier wave has an r.m.s. value of 10 V when it is unmodulated and of 10.8 V when it is sinusoidally modulated. Calculate its depth of modulation. The carrier is then modulated by a signal that contains components at two different frequencies and its r.m.s. value goes up to 10.8 V. If the depth of modulation due to one frequency is 40% calculate the depth of modulation produced by the other frequency. What is the overall depth of modulation?

1.3 A 1 MHz carrier wave is amplitude modulated to a depth of 80% by a 5 kHz sinusoidal signal and is then applied to a single tuned circuit that is resonant at 1 MHz and of Q-factor 100. Calculate the depth of modulation of the output signal.

1.4 A carrier $12 \sin (8 \times 10^6 \pi t)$ volts is amplitude modulated by the signal $v = 6 \sin (2000 \pi t) + 3 \cos (4000 \pi t)$ volts. Calculate the depth of modulation. If the modulated wave is applied across a 100 Ω resistance calculate the power dissipated.

1.5 A tuned circuit has a Q-factor of 60, a capacitance of 100 pF and is resonant at 455 kHz. It is effectively connected in parallel with the input resistance of a diode detector having a load resistance of 250 kΩ. If the detection efficiency is 86% calculate the effective Q-factor of the tuned circuit.

1.6 A carrier wave is amplitude modulated to a depth of 60%. One of its sidebands is then completely suppressed. Determine the percentage second-harmonic distortion of the modulation envelope.

1.7 Figure Q.1 shows the circuit of an amplitude modulator whose output tuned circuit is resonant at 10^7 rad/s and has an effective Q-factor of 50. The carrier frequency is 10^7 rad/s and the carrier voltage is large enough to switch T_1 and T_2 alternately ON and OFF. Calculate (a) the d.c. current of T_2, (b) the amplitude of the unmodulated carrier output, and (c) the peak audio input voltage which will give a modulation depth of 100%. For all transistors $h_{FE} = 100$, $V_{BE} = 0.65$ V.

1.8 Show that the improvement in system signal-to-noise ratio obtained by converting a d.s.b. system to s.s.b.s.c. working is given by $3 + 20 \log_{10} [(1 + m)/m]$ dB.

The signal-to-noise ratio at the output of a radio system is 30 dB when an 80% modulated d.s.b. 10 kW transmitter is used. Calculate the output signal-to-noise ratio when the system is converted to operate as s.s.b.s.c. with a transmitted power of 4 kW.

1.9 A sinusoidal signal has an r.m.s. value of 10 V before it is amplitude modulated and 10.4 V after modulation. The modulated signal is passed through a network that attenuates the lower side-frequency by 10 dB but leaves the carrier and upper side-frequency unchanged. Calculate the r.m.s. value of the wave.

1.10 Show that a square-law device can be used for either amplitude modulation or demodulation.

Chapter 2

2.1 The r.f. power output of a radio transmitter is 1 kW when it is not modulated. Calculate the power output when

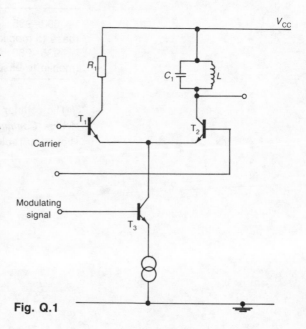

Fig. Q.1

the transmitter is modulated (*a*) in amplitude with a depth of modulation of 50%, and (*b*) in frequency with a modulation index of 0.5. For each case draw frequency spectrum diagrams showing the relative amplitudes of the components.

2.2 A single-channel a.m. radio-telephony system has an output signal-to-noise ratio of 30 dB. If the system is changed to frequency modulation with a rated system deviation of 10 kHz and the transmitted power is doubled, what will then be the output signal-to-noise ratio? The maximum modulating frequency is 3 kHz.

2.3 Explain the effect that pre-emphasis and de-emphasis have on the output signal-to-noise ratio of an f.m. system. If the de-emphasis network has a time constant of 75 μs, calculate the improvement in signal-to-noise ratio for a baseband signal having a maximum frequency of 15 kHz and a rated system deviation of 75 kHz.

2.4 Derive an expression for the instantaneous voltage of a 100 MHz carrier that is frequency modulated by a 5 kHz 5 V tone if a signal voltage of 10 V produces the rated system deviation. Use Table 2.1 to determine the amplitudes of the various components in the f.m. wave and then find the percentage power in the first ($m + 1$) side-frequencies.

2.5 Explain what is meant by the triangular noise spectrum of an f.m. system. A carrier of 1.5 mV at 100 MHz is received together with an interfering signal of 50 μV at 100.1 MHz. Calculate the peak frequency deviation of the carrier.

2.6 A varactor diode modulator operates with a reverse-bias voltage of 8 V and has a capacitance of 30 pF. The oscillator has a tuning capacitor of 60 pF and oscillates at 6 MHz. Calculate the peak value of the modulating signal voltage required to deviate the carrier frequency by 16 kHz.

2.7 Discuss the relative merits of frequency and phase modulation and explain how they differ from one another.

Draw spectrum diagrams for a carrier wave (*a*) frequency-modulated by a 2 MHz signal with a rated system deviation of 4.8 MHz, and (*b*) phase-modulated by a 2 MHz signal with a peak phase deviation of 2.4 rad.

2.8 The output of a phase modulator is a carrier of 1 MHz and a frequency deviation of 1 kHz. The modulated signal is passed through three frequency multipliers in turn that have, respectively, multiplication factors of 6, 8 and 6. The signal is then applied to a mixer together with a 200 MHz tone. Calculate the centre frequency and the deviation of the output of the mixer. Deduce a combination of multipliers and mixers that will turn the modulator output into a 96 MHz carrier with 75 kHz frequency deviation.

Chapter 3

3.1 An r.f. line is $\lambda/2$ long, has a characteristic impedance of 600 Ω, 6 dB loss and it is terminated by a load impedance of $1200 + 300$ Ω. Calculate the input impedance of the line.

3.2 List the factors that limit the use of inductors and capacitors at very high frequencies. Calculate the length of loss-free line that would simulate an inductance of 20 nH at 600 MHz. $Z_0 = 50$ Ω.

3.3 A $\lambda/8$ length of loss-free line has a characteristic impedance of 600 Ω and a load of 800 Ω. Calculate the input impedance of the line. Also calculate the ratio, in dB, of the powers in the load with and without the insertion of a $\lambda/4$ section of loss-free matching line of the appropriate impedance.

3.4 A 75 Ω transmitting aerial, operating at 500 MHz, is connected to the transmitter by a 6 m length of coaxial cable of 75 Ω characteristic impedance. A cable fault causes an effective capacitance of 4.244 pF to be in series with the line at a distance of 2 m from the aerial. Calculate the value of the voltage-reflection coefficient on the cable (*a*) at the fault, and (*b*) at the input to the line. Also calculate the power delivered to the aerial before and after the fault occurs if the transmitter power output is 30 W.

3.5 A load of $210 - j180$ Ω is connected to the output terminals of a 150 Ω loss-free line at 100 MHz. Find the position and the length of a single matching stub made of the same cable. Calculate the voltage across the resistive part of the load if the sending-end voltage is 100 μV.

3.6 An unknown load presents an impedance of $80 + j110$ Ω when supplied via a line of characteristic impedance 100 Ω and length 0.154λ. Use a Smith chart to find (*a*) the load impedance; (*b*) the v.s.w.r., (*c*) the shortest distance from the load at which a single matching stub could be connected, and (*d*) the length of this stub.

3.7 A load of $100 - j100$ Ω is to be matched to a 50 Ω feeder by connecting a short-circuit stub across the line 0.1λ from the load, and then using a $\lambda/4$ section to connect this point to the rest of the feeder. Use a Smith chart to calculate the length of the stub and the impedance of the $\lambda/4$ section.

3.8 A coaxial cable has an attenuation of 0.2 dB per metre. Calculate the Q-factor of a piece of this cable that is resonated at a frequency of 900 MHz.

3.9 A loss-free line divides into two sections A and B. The length of section A is 10 cm and that of section B is 14.5 cm. The impedances of the loads connected to the two sections of line are $166 + j60$ Ω on to A and $37.5 + j0$ Ω on to B. The frequency of operation is 258 MHz and the characteristic impedance is 100 Ω. Calculate the voltage-reflection coefficient on each line and also the v.s.w.r. on the source side of the junction.

3.10 A 50 Ω slotted line is used to measure the impedance of an unknown load. With the load connected the v.s.w.r. on the line was $S = 2.0$. Adjacent voltage minima were found to be at 41.2 cm and 71.2 cm from the load. With the load disconnected and replaced by a short-circuit the voltage minima were found to be at 33.7 cm and 63.7 cm from the load. Use a Smith chart to calculate the impedance of the load.

Chapter 4

4.1 Explain why a transverse electromagnetic wave cannot be propagated down a rectangular waveguide. Show how a TE wave can be regarded as being the resultant of two TEM waves. A waveguide of dimensions 2.8 cm by 1.3 cm has a group wavelength equal to the cut-off wavelength. Calculate the frequency of the propagating signal.

4.2 Explain why the cut-off frequency is an important parameter of a waveguide. Also explain why the power-handling capability of a waveguide is limited by dielectric breakdown. A waveguide operating at 3.2 GHz has a breakdown electric field strength of 3×10^6 V/m. If a safety factor of 2 is used calculate the maximum power that can be transmitted.

4.3 A rectangular waveguide has internal dimensions of 0.569 cm and 0.285 cm. Calculate (*a*) the cut-off frequency, (*b*) the guide wavelength, and (*c*) the phase and group velocities if the frequency is 40 GHz.

4.4 A rectangular waveguide has internal dimensions of 1.067 cm and 0.4318 cm. It is connected to another similar waveguide that is filled with a dielectric of relative permittivity 2. This second waveguide is matched to its load. If the frequency of operation is 20 GHz calculate the v.s.w.r. in the air-filled waveguide.

4.5 What is meant by the term 'dominant mode' in a rectangular waveguide? What are higher modes and why are they generally undesirable? How can the higher-order modes be suppressed? A rectangular waveguide has dimensions 1.580 cm and 0.7899 cm and transmits a signal whose frequency is twice the cut-off frequency. Calculate the angle at which reflection from the walls occurs.

4.6 In a test on a 31 cm length of mismatched waveguide the measured v.s.w.r. pattern has adjacent minima of 2.9 cm and the normalized input impedance was $3 - j2 \ \Omega$. A matching iris having an admittance of $-j1.43$ S is then connected in parallel with the mismatched load. Use a Smith chart to find the v.s.w.r. on the waveguide.

4.7 A rectangular waveguide has internal dimensions of 1 cm and 2.3 cm and it is operated at 10 GHz. Calculate (*a*) the free-space wavelength, (*b*) the cut-off frequency, (*c*) the phase velocity, (*d*) the group velocity, and (*e*) the guide wavelength.

4.8 A rectangular waveguide has internal dimensions of 2.8 cm and 1.3 cm and it is operated at a frequency equal to twice the cut-off frequency of the dominant mode. Calculate (*a*) the guide wavelength, and (*b*) the angle at which reflection from the waveguide wall takes place.

4.9 A rectangular waveguide has wide dimension *a* of 2.3 cm. A wave travelling down this guide makes an angle of reflection of 60° at each wall. Calculate (*a*) the phase velocity, (*b*) the group velocity, (*c*) the frequency of operation, (*d*) the guide wavelength, and (*e*) the cut-off wavelength.

Chapter 5

5.1 Define the terms 'noise factor' and 'noise temperature' and derive the relationship between them. Figure Q.2 shows a radio system. Calculate the required available signal power delivered by the aerial for the output signal-to-noise ratio to be better than 30 dB. The bandwidth is 10 MHz.

5.2 Explain how galactic noise and atmospheric noise vary with frequency. Hence explain why certain frequencies are used for communications satellite systems.

An aerial is connected to a radio receiver by a feeder of 1.46 dB loss, and a low-noise amplifier of 20 dB gain and a noise temperature of 89.9 K. If the noise factor of the radio receiver is 4.0 dB calculate (*a*) the overall noise factor of the feeder, amplifier and receiver, and (*b*) the system noise temperature.

5.3 A radio receiver has a noise factor of 5 dB and it is connected to an aerial of noise temperature 133 K by a feeder of (*a*) 0 dB loss, and (*b*) 1.76 dB loss. Calculate the output noise power if the bandwidth is 10 MHz, and the gain of the receiver is 60 dB.

5.4 An amplifier has a noise factor of 7 dB and its output terminals are connected to a radio receiver whose noise factor is 10 dB. Calculate the minimum gain the amplifier must have for the system noise temperature to be less than 1500 K.

5.5 Three amplifiers have the parameters shown in Table Q.1. Determine the order in which the amplifiers ought to be connected in order to give the lowest overall noise factor. What is the value of this minimum noise factor?

Fig. Q.2

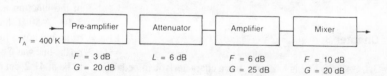

$T_A = 400$ K

Pre-amplifier	Attenuator	Amplifier	Mixer
$F = 3$ dB	$L = 6$ dB	$F = 6$ dB	$F = 10$ dB
$G = 20$ dB		$G = 25$ dB	$G = 20$ dB

Table Q.1

Amplifier	Gain (dB)	Noise factor (dB)
1	20	7
2	10	5
3	6	2

5.6 An aerial is connected to an amplifier that has a gain of 26 dB and a noise factor of 3 dB. The noise temperature of the aerial is 300 K and it delivers a signal power of 1 nW to the amplifier. If the available noise power from the aerial is −110 dBm calculate (a) the input signal-to-noise ratio, (b) the output signal-to-noise ratio.

5.7 An aerial whose noise temperature is 20 K is connected by a waveguide feeder of 0.3 dB loss to a low-noise amplifier of noise temperature 10 K and gain 28 dB. The output of the low-noise amplifier is connected to a TWT that has a noise factor of 10 dB and a gain of 25 dB. If the noise bandwidth of the system is 2.8 MHz, calculate what signal power must be delivered by the aerial to give an output signal-to-noise ratio of 30 dB.

5.8 The noise generated within an r.f. amplifier can be assumed to be generated in an equivalent resistance of 56 kΩ across the input terminals. The noise temperature of this resistance is 290 K and the bandwidth of the amplifier is 10 kHz. Calculate the output signal-to-noise ratio of the amplifier when the input signal is a 100 μV voltage sinusoidally modulated 30% at 1 kHz.

5.9 Explain why the first stage of a radio receiver should have a low noise factor and a high gain. Why, for the latter stages, may gain be more important than noise factor? A radio receiver has a noise factor of 9 dB and a bandwidth of 100 kHz. Calculate the input signal power needed to give an output signal-to-noise ratio of 20 dB.

5.10 (a) Show that the r.m.s. noise voltage generated in the parallel combination of a resistor R and a capacitor C is given by $\sqrt{(kT/C)}$ V. (b) Calculate the variation in the noise temperature of a circuit as its noise factor is varied from 0 dB to 2 dB.

Chapter 6

6.1 Show that the radiation resistance of an earthed $\lambda/4$ vertical aerial is approximately 40 Ω.

An aerial of this type is supplied with a current of 10 A r.m.s. at 4 MHz. Calculate (a) the effective height of the aerial, (b) the electric field strength produced at a point at ground level 30 km away, (c) the power received by an aerial of gain 12 dBi, and (d) the transmitted power.

6.2 Explain the meanings of the terms 'gain', 'effective aperture' and 'directivity' as applied to an aerial. Two aerials that operate at 600 MHz have effective apertures of 4 m and 2 m, respectively. Calculate the gain of each aerial with respect to (a) an isotropic radiator, and (b) a $\lambda/2$ dipole.

6.3 What is meant by the terms the 'induction field' and the 'radiation field' of an aerial? At what distance from an aerial operating at 3 MHz are the two fields of equal amplitude? If, at a much greater distance, the radiation field has a magnetic field strength of 265×10^{-6} AT/m what is (a) the electric field strength, and (b) the power density at this point?

6.4 100 kW power is radiated by an aerial whose effective length is 100 m. Calculate the field strength produced at a distance of 100 km if the frequency is 60 kHz.

6.5 Calculate the electric field strength at ground level at a point 10 km from a $\lambda/2$ monopole. The aerial is supplied with a 3 A r.m.s. current at a frequency of 10 MHz.

6.6 Explain what is meant by the effective height of an aerial. An earthed monopole has an effective height of 0.1λ and it is supplied with an r.m.s. current of 100 A. Calculate the total radiated power and the effective radiated power. Also find the field strength at ground level 50 km distant.

6.7 The transmitting aerial shown in Fig. Q.3 is supplied with current at 80 A peak and at frequency 666.66 kHz. Calculate (a) the effective height of the aerial, and (b) the electric field strength produced at ground level 40 km away.

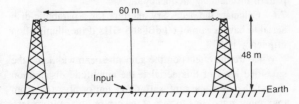

Fig. Q.3

6.8 Show that the power received by an aerial of gain G_r is given by $P_r = P_t G_t G_r (\lambda/4\pi D)^2$. Calculate the total attenuation if $G_t = G_r = 30$ dBi, $D = 30$ km, and the frequency of operation is 600 MHz.

6.9 Calculate the radiation resistance of a $\lambda/16$ aerial. Determine its radiated power when the input current is 100 A r.m.s. If the loss resistance is 25 Ω calculate the efficiency of the aerial.

6.10 An aerial is 100 m in height and carries a current at 100 kHz whose amplitude decreases linearly from 200 A at the base to zero at the top. Calculate the effective height of the aerial when it is mounted upon perfectly conducting earth. Calculate the field strength produced at a distance of 20 km from the aerial.

Chapter 7

7.1 Two vertical $\lambda/2$ dipoles are in the same horizontal plane and spaced $\lambda/4$ apart. The dipoles are supplied with currents of $I\angle 0°$ and $1.5I\angle 90°$, respectively. Calculate and plot the horizontal-plane radiation pattern of the array.

7.2 An aerial array consists of four vertical $\lambda/2$ dipoles in the same horizontal plane spaced $\lambda/2$ apart. The currents fed to the aerials are of the same magnitude but have a progressive phase difference of $90°$. Derive the expression for the horizontal-plane radiation pattern of the array and show that maximum field strength occurs at $120°$ to the line of the array.

7.3 A Yagi aerial consists of a dipole of $73 + j0\ \Omega$ impedance and a parasitic element of $81 + j108\ \Omega$ impedance. The mutual impedance between the two elements is $40 - j30\ \Omega$. Calculate the ratio of the powers radiated in the directions dipole to parasitic and parasitic to dipole. Hence state whether the parasitic element is a reflector or a director.

7.4 An aerial array consists of three $\lambda/2$ dipoles A, B and C that are fed with equal-amplitude, in-phase currents. Each dipole has a radiation resistance of $73\ \Omega$ and their mutual impedances are $Z_{AB} = Z_{BC} = -12.5 - j30\ \Omega$, $Z_{AC} = 5 + j17.5\ \Omega$. Calculate the gain of the array.

7.5 Draw the radiation pattern of an h.f. log-periodic aerial for both the horizontal and vertical planes. Why is the l.p.a. generally preferred for long-distance radio h.f. systems? Draw a sketch of an l.p.a. suitable for use at about 5 to 20 MHz. What is the effect on the main lobe of the radiation pattern of changing frequency?

7.6 Calculate the necessary aperture for a parabolic-dish aerial to have a gain of 60 dBi at 5 GHz if the illumination efficiency is 60%.

What are the effects on the gain, the beamwidth and the side-lobe level of this aerial if the amplitude distribution across the dish aperture varies (a) uniformly, (b) from maximum at the edge to zero at the centre, or (c) from zero at the edge to maximum at the centre?

7.7 What is meant by the principle of pattern multiplication?

An aerial array consists of 15 vertical $\lambda/2$ dipoles arranged in three rows of five dipoles. Derive an expression for the horizontal-plane radiation pattern.

7.8 What is meant by the gain of a transmitting aerial? An aerial is operated at 11 GHz; calculate its effective aperture if its gain is 56 dBi. Explain what is meant by the term 'beamwidth' and calculate its value for this aerial. Why are side lobes undesirable in (a) a transmitting aerial, and (b) a receiving aerial?

7.9 Three vertical $\lambda/2$ dipoles are mounted 3λ apart in the same horizontal line and are fed with currents of $I_A = 0.25I \sin(\omega t - 90°)$, $I_B = I \sin \omega t$ and $I_C = 0.25I \sin(\omega t + 90°)$.

Derive an expression for the field strength produced at a distant point.

7.10 Show that the vertical-plane radiation pattern of a horizontal $\lambda/2$ dipole mounted at a height of h above the ground will have maxima at angles of elevation of $\sin^{-1}(\lambda/4h)$, $\sin^{-1}(3\lambda/4h)$, $\sin^{-1}(5\lambda/4h)$, $\sin^{-1}(7\lambda/4h)$, etc. Sketch the radiation patterns if $h = \lambda$, 2λ, and 3λ.

Chapter 8

8.1 A 2 GHz radio link operates over a 48 km stretch of sea with aerials at equal heights. Calculate the necessary heights of the aerials if the k-factor is 0.65 and the direct wave is to clear the sea by a distance equal to 0.577 times the first Fresnel zone. If a second aerial is fitted to give height diversity, calculate its height also.

8.2 An ionospheric layer has a maximum electron density of 1.6×10^{11} electrons/m^3 and it is at a virtual height of 162 km. Calculate its m.u.f. if the radius of the earth is 6400 km and the sky-wave link is 2000 km long.

8.3 At a distance of 1 km from a 1 MHz radio transmitter the daytime field strength is 200 mV/m. Calculate the field strength 110 km from the transmitter if the ground-wave attenuation is 9 dB greater than the free-space loss. Also calculate the field strength at the same point during the night when there is also a sky wave received. Assume the sky wave to be reflected from a height of 120 km and that the earth is flat.

8.4 Explain how the field strength of a u.h.f. transmitter varies with the distance from the transmitter.

Calculate the maximum range for a u.h.f. radio link if the transmitting aerial is at a height of 110 m and the receive aerial height is 60 m. Take the radius of the earth as 6400 km.

8.5 A transmitting aerial has a gain of 6 dBi and is mounted 180 m above flat earth. The transmitted power is 5 W at 50 MHz and the 16 m high receive aerial is 18 km distant. Calculate the total field strength at the receive aerial if the reflection coefficient of the earth is -0.8.

8.6 A communications satellite is in orbit 35 800 km above an earth station. The down path from satellite to earth station operates at 4 GHz. At the earth station the receiver has an effective input noise temperature of 60 K and a bandwidth of 30 MHz, the parabolic dish aerial has a gain of 60 dBi, and the aerial noise temperature is 40 K. If the e.r.p. of the satellite is 30 dBW, calculate (a) the carrier-to-noise ratio of the receiver, and (b) the G/T ratio.

8.7 A communications satellite link has the following data: operating frequency 6 GHz, gain of satellite aerial 6 dBi, gain of earth station aerial 50 dBi, noise temperature of earth station aerial 290 K, noise factor of receiver 3 dB, bandwidth of receiver 20 MHz, and link length 36 000 km.

The carrier-to-noise ratio at the earth station is to be 30 dB. Calculate (a) the minimum signal power density at the earth station receive aerial, (b) the minimum field strength at the receive aerial, and (c) the minimum power that must be transmitted by the satellite.

8.8 Discuss how the propagation of h.f. radio waves is affected by sun spots.

The maximum electron density in the ionospheric layer is 1.6×10^{11} electrons/m^3 and occurs at a height of 162 km. Calculate the skip distance if the radius of the earth is 6400 km and the m.u.f. is 5.6 MHz.

8.9 Explain briefly the way in which a tropospheric scatter system works.

A 2.5 GHz tropospheric scatter system has its terminals 200 km apart and uses aerials whose diameters are both 6.1 m. If the transmitted power is 20 kW and the over-the-horizon loss is 62 dB calculate the signal power supplied to the receiver.

8.10 What is meant by the term 'Fresnel zone' and what is the use of this zone in radio-communications?

A 600 MHz signal is transmitted from a 75 m high aerial towards the receive aerial 30 km away. There is a 42 m high obstacle 20 m from the transmitter which must be cleared by the direct ray by a distance equal to 0.577 times the radius of the first Fresnel zone. Assuming the k-factor to be 0.7, determine the minimum height for the receive aerial.

Chapter 9

9.1 An amplifier has a gain of 15 dB and has a two-tone input signal (each tone at -15 dBm) applied to its terminals. The third-order intercept point is $+22$ dBm. Calculate the magnitude of the third-order intermodulation products.

9.2 A radio receiver has a gain of 20 dB and a third-order intercept point of $+20$ dBm. The input signal consists of two tones, one at 0 dBm and the other at -12 dBm. The third-order intermodulation products are at -12 dBm. Calculate the third-order intercept point. Also calculate the noise floor if the bandwidth is 3000 Hz and the noise factor is 8 dB.

9.3 Explain why the noise factor of a radio receiver is an important feature at v.h.f. and u.h.f. but not at h.f.

A v.h.f. radio receiver has a noise factor of 9 dB, an a.f. bandwidth of 3 kHz and an output signal-to-noise ratio of 25 dB. Calculate (a) the input signal-to-noise ratio, and (b) the sensitivity of the receiver.

9.4 (a) Explain, with reference to a modern communication receiver, why the double superheterodyne principle is often used. (b) What is meant by the selectivity of a radio receiver and how is it provided? (c) Explain what is meant by reciprocal mixing and how it affects the selectivity of a radio receiver.

9.5 Draw the block diagram of a double superheterodyne radio receiver that can tune over the frequency range 3 to 30 MHz. State typical figures for the first and second intermediate frequencies, and label the frequencies and typical voltages of signals at each block in the diagram if the receiver is tuned to 10 MHz.

9.6 Why is a.g.c. always applied to an amplitude-modulation radio receiver but not always to a frequency-modulation receiver? Explain the problems associated with the application of a.g.c. to (a) a d.s.b.a.m. receiver, and (b) an s.s.b. receiver. How may the gain of the front-end be controlled by a.g.c. and what are the relative merits of the methods?

9.7 A mixer has the input–output voltage characteristic given by $v = 2 + 1.4v + 0.8v^2 + 0.4v^3 + \ldots$ and has two 0 dBm signals at different frequencies applied to it. Calculate the magnitude of the second- and third-order intermodulation products. Plot the characteristic for the wanted signal and both the second- and third-order intermodulation products and hence determine the second- and third-order intercept points.

9.8 Briefly explain the function of each block in the radio-receiver circuit given in Fig. 9.15.

Chapter 10

10.1 A 600 channel microwave system uses receivers with an i.f. bandwidth of 30 MHz, a frequency deviation of 200 kHz, and a maximum channel frequency of 2.54 MHz. Calculate the carrier-to-noise ratio corresponding to a weighted output signal-to-noise ratio of 71.58 dB.

10.2 A digital radio-relay link operates at 11.2 GHz and is 45 km long. Atmospheric losses are equal to 1.5 dB. The transmitted power is 10 W with 9 dB feeder losses and both the transmit and receive aerials have a gain of 49 dB. The receiver has an i.f. bandwidth of 80 MHz, a noise factor of 8 dB and feeder losses are 8 dB. The minimum carrier-to-noise ratio required to obtain a b.e.r. of 1×10^{-6} is 18 dB. Calculate the fade margin.

10.3 List all the important sources of noise in an analogue radio-relay link carrying 960 channels. For each source state whether the source is thermal or intermodulation in its nature.

10.4 Discuss the effects of transmission delay on telephone signals routed via a communications satellite system.

Two earth stations are 9650 km apart and are equidistant from a satellite that is in orbit at a height of 35 880 km. The up-link operates at 6 GHz and the down-link at 4 GHz. If the satellite introduces a gain of 113 dB and the earth-station aerials are each 16.4 m in diameter calculate the overall path loss, and the transmission delay.

10.5 Draw the block diagram of the equipment used in the terminal stations of a terrestrial radio-relay link. Explain the function of each block.

10.6 Describe the operation of an INTELSAT system with particular reference to the orbit used, the carrier frequencies and the on-board processing. Explain why cross polarization is often employed. Why is it that earth stations are located well remote from large towns or cities even though most of the traffic handled is destined for the town or city?

10.7 An analogue 960 channel radio-relay system operates at 6 GHz. The transmitter output power is 5 W, feeder losses are 5 dB, and both aerials have a gain of 41.5 dB. The receiver has a noise factor of 10 dB, an i.f. bandwidth of 40 MHz and is 56 km distant. State the r.m.s. frequency deviation due to the test tone and then calculate (*a*) the power received by the aerial, (*b*) the carrier-to-noise ratio, and (*c*) the output signal-to-noise ratio in the top channel.

10.8 Explain the principle of white-noise testing of a frequency-division multiplex radio-relay system. State the meaning of the term 'noise power ratio'. In a 60 channel system the test-tone level is set to be -12 dBm in each channel at the input of the link. Calculate the noise power required to simulate busy-hour loading.

10.9 Explain the meanings of the terms 'white noise testing' and 'noise power ratio'. Calculate the r.m.s. frequency deviation due to the white noise signal in a 960 channel system. Also calculate the i.f. bandwidth necessary.

Answers to Numerical Exercises

1.1 64%, 4.77 A
1.2 57.7%, 41.6%, 57.7%
1.3 56.6%
1.4 55.9%, 832.5 mW
1.5 24.55
1.6 7.5%
1.7 (a) 1.05 mA, (b) 6.4 V, (c) 4 V
1.8 36 dB
1.9 10.22 V

2.1 (a) 1125 W, (b) 1000 W
2.2 48.23 dB
2.3 6.26 dB
2.5 3333 Hz
2.6 0.25 V
2.8 488 MHz, 288 kHz

3.1 716 Ω
3.2 7.85 cm
3.3 600 ∠ −30° Ω, 0.09 dB
3.4 (a) 0.45 ∠ −63°, (b) 0.45 ∠ −178°,
30 W, 23.93 W
3.5 0.307 m, 0.357 m, 118.2 μV
3.6 (a) 30 Ω, (b) 3.2, (c) 0.08λ, (d) 0.40λ
3.7 0.335λ, 26.5 Ω
3.8 409
3.9 0.285 −j0161, −456, 1.7
3.10 25 + j0 Ω

4.1 7.576 GHz
4.2 5.57 MW
4.3 26.36 GHz, 1.151 cm, 4.604×10^9 m/s, 19.55×10^6 m/s
4.4 1.4
4.5 29.8°
4.6 2.4
4.7 (a) 3 cm, (b) 6.52 GHz, (c) 3.957×10^8 m/s,
(d) 2.274×10^8 m/s, (e) 3.957 cm
4.8 (a) 3.23 cm, (b) 30°
4.9 (a) 6×10^8 m/s, (b) 1.5×10^8 m/s,
(c) 13.04 GHz, (d) 2.65 cm, (e) 4.6 cm

5.1 101 pW
5.2 (a) 2.68, (b) 248 K
5.3 (a) 105 nW, (b) 112 nW
5.4 8.85 dB
5.5 3,2,1; 3.47 dB
5.6 50 dB, 47.1 dB
5.7 2.17×10^{-15} W
5.8 20 dB
5.9 3.18×10^{-13} W
5.10 (b) 0 K to 169.6 K

6.1 (a) 11.94 m, (b) 20 mV/m, (c) 7.5 mW, (d) 757 W
6.2 (a) 23 dBi, 20 dBi, (b) 20.85 dB, 17.85 dB
6.3 (a) 100 mV/m, (b) 26.5 μW/m^2
6.4 30 mV/m
6.5 18 mV/m
6.6 157.9 kW, 473.7 kW, 75.4 mV/m
6.7 (a) 42.46 m, (b) 71.1 mV/m
6.8 57.55 dB
6.9 1.54 Ω, 15.4 kW, 5.8%
6.10 100 m, 126 mV/m

7.3 7.2 dB. Director
7.4 7.13 dB
7.6 478 m^2
7.8 23.56 m^2, 3.5°

8.1 93.71 m, 113 m
8.2 15.36 MHz
8.3 0.23 mV/m, 0.53 to 0.99 mV/m
8.4 75.3 km
8.5 0.49 mV/m
8.6 28.27 dB, 40 dB K^{-1}
8.7 (a) 8×10^{-12} W/m^2, (b) 55 μV/m, (c) 32.9 kW
8.8 453 km
8.9 14.6 nW
8.10 109 m

9.1 −44 dBm
9.2 2 dBm, −131.2 dBm
9.3 (a) 34 dB, (b) 2.44 μV
9.7 −32 dBm, −70.5 dBm, +18 dBm, +26 dBm

10.1 47.3 dB
10.2 42 dB
10.4 170 dB, 0.25 s
10.7 (a) 1.6 W, (b) 60 dB, (c) 81 dB
10.8 −16 dBm
10.9 1102 MHz, 15.34 MHz

Index